Advances in Underwater Technology, Ocean Science and Offshore Engineering

Volume 11

Underwater Construction: Development and Potential

ADVANCES IN UNDERWATER TECHNOLOGY, OCEAN SCIENCE AND OFFSHORE ENGINEERING

Vol. 1 Developments in Diving Technology
Vol. 2 Design and Installation of Subsea Systems
Vol. 3 Offshore Site Investigation
Vol. 4 Evaluation, Comparison and Calibration of Oceanographic Instruments
Vol. 5 Submersible Technology
Vol. 6 Oceanology
Vol. 7 Subsea Control and Data Acquisition
Vol. 8 Exclusive Economic Zones
Vol. 9 Stationing and Stability of Semi-submersibles
Vol. 10 Modular Subsea Production Systems
Vol. 11 Underwater Construction: Development and Potential

CONFERENCE PROGRAMME PLANNING COMMITTEE

R.W. Barrett *Underwater Engineering Group*
C. Baxter *Advanced Production Technology Ltd*
J. Pritchard *Society for Underwater Technology (Conference Organiser)*
A.C. Walker *University of Surrey*

Advances in Underwater Technology, Ocean Science and Offshore Engineering

Volume 11

Underwater Construction: Development and Potential

Proceedings of an international conference (The Market for Underwater Construction) organized by the Society for Underwater Technology and held in London, 5 & 6 March 1987.

Published by
Graham & Trotman
A Member of the Kluwer Academic Publishers Group

First published in 1987 by

Graham & Trotman Limited
Sterling House
66 Wilton Road
London SW1V 1DE
UK

Graham & Trotman Inc.
101 Philip Drive
Assinippi Park
Norwell, MA 02061
USA

ISBN-13: 978-94-010-7953-2 e-ISBN-13: 978-94-009-3239-5
DOI: 10.1007/978-94-009-3239-5

LCCCN 87-17709

Library of Congress Cataloging-in-Publication Data

Underwater construction.

(Advances in underwater technolygy, ocean science, and offshore engineering; v. 11)
Includes bibliographies.
1. Underwater construction – Congresses. I. Society for Underwater Technology. II. Series.
TC195.U55 1987 627'.702 87-17709

Typeset by Bookworm Typesetting, Manchester

Contents

Introduction

Falling oil prices and smaller offshore fields, especially in the UK sector of the North Sea, have produced a resurgence of interest in subsea developments. These developments always include the installation of a subsea structure and laying and tying-in of pipeline and control lines. In the Southern North Sea small unmanned jackets may become widely used and these require subsea control and power lines installation.

This change in the offshore scene has produced a potentially larger market for underwater construction activities. Companies engaged in underwater design and construction have been developing new equipment and techniques to enable further economies to be realised.

It is not only in hydrocarbon development where underwater construction plays a major role. One of the largest offshore construction projects on the UK Continental Shelf in recent years is the cross-channel link where power cables have been laid between the UK and France.

This volume looks at the economic outlook and the breadth of underwater construction operations; important developments in techniques and equipment are presented together with a discussion of various projects in which they have been successfully used.

PART I

1

An overview of subsea construction

R. Goodfellow, Goodfellow Associates Ltd

INTRODUCTION

Working underwater is a lot more difficult than working above water or on land, therefore the incentives to do so must be found in some aspects of project realization, such as:

- reduced cost;
- advantageous schedule;
- improved technology.

In this chapter these aspects will be considered for a number of cases of floating systems and subsea developments which have been carried out. The North Sea provides examples of floating systems, starting with Hamilton Brothers' Argyll, and the development of the buoyant tension-legged platform of Conoco Hutton TLP in production in 1984.

Subsea systems are complementary to existing fixed structures and increase the recoverability; for example, BP Magnus has seven satellite wells.

Advances in Underwater Technology, Ocean Science and Offshore Engineering, Volume II: Underwater Construction: Development and Potential

Subsea production began with the Zakum Subsea Production Scheme (1969–72) and now subsea systems are used in field developments in the North Sea; for example, the Shell UMC, Texaco Highlander and North-East Frigg.

This chapter also discusses some prospective developments in applications, techniques and equipment. Other chapters in this volume will elaborate on many of the topics which will be raised.

Early practitioners of the art of working underwater include marine salvagers and offshore drillers, and techniques which were developed for these activities are being elaborated and improved upon for those more complex activities of offshore oilfield development. Experience, and common sense, has taught that complicated construction work should be done as completely as possible on land before taking units offshore for installation, and that only minimal assembly of pieces be required subsea.

Examples of subsea construction activities include well templates, manifolds, pipelines, multiple flowlines, anchor foundations, use of divers and remotely operated vehicles. Prospects for the future of subsea activities are assured by the present keen interest in floating production systems and subsea satellite developments to be tied-back to existing installations.

The future North Sea development in the UK sector will be primarily:

- smaller fields;
- satellite fields;
- condensate/sour gas.

Deepwater developments present greater challenges but, with the discovery of sufficient reserves, there is no doubt that these developments will also stimulate new requirements as well as new capabilities for underwater construction.

ARGYLL FIELD

The Argyll Field layout is represented in Fig. 1, and the key information summary is presented in Table I. The production riser system is of particular interest.

Production riser system

The production riser system at Argyll is made up of standard drilling components assembled in what was a unique system, in 1975. The

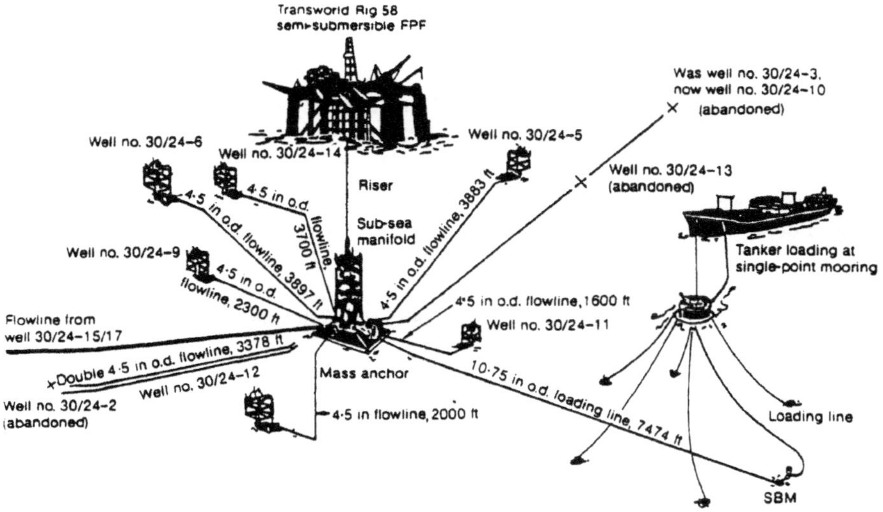

Fig. 1 Argyll Field general layout.

system consists of five basic elements from the sea bed upwards – the mass anchor, permanent base, manifold, risers and flexible connections.

The central riser is of 10 in. * nominal bore and serves as the main supporting member and export (or shipping) riser. This element consists of a stab sub-assembly, universal joint and six 40 ft* joints of 10 in. riser pipe connected with standard marine riser joints. There are eight production risers positioned radially around the 10 in. riser, each of 4 in. nominal bore, made up of a stab sub, safety joint, one 6 in. × 40 ft riser joint and five 40 ft joints of 4 in. riser pipe connected by 4 ½ in. internal-fitting tool joints. Lateral movement of the floating production facility is transmitted to the top of the riser assembly, which deflects to accommodate the movement. The 10 in. riser accommodates this movement via the universal joint and the 4 in. risers bend to follow this deflection. Also the larger diameter and increased wall thickness of the bottom pipe joint contributes to the strength to tolerate the continuous flexing forces.

Positioning of the 4 in. risers is maintained by guide funnels located radially and at regular intervals over the length of the 10 in. riser. The radius of bending of the 4 in. risers around the 10 in. universal joint is maximized by the absence of guide funnels over some 98 ft of the lower ends of the riser assembly.

Goosenecks terminate the upper ends of each of the risers in the semi-submersible/moonpool; these connect to the fixed pipework of

*1 inch ≈ 2.54 cm; 1 ft ≈ 30.5 cm.

Table I
Argyll Field – key information summary

Operator, partners	Hamilton Brothers, RTZ, Texaco, Blackfriars Oil, Trans-European Co.
Discovery date	October 1971
Water depth	79 m (260 ft)
Sea-bed nature	
Reservoir depth	2743 m (9052 ft) Zechstein & Rotliegendes
Recoverable reserves	
Original	75 million barrels
Remaining	35 million barrels (31 December 1981)
Recovery factor 35%	API gravity 36° GOR 210 scf/barrel Sulphur content 0.2%
Production start	June 1975
Peak production	23 000 barrels/day in 1976
1981 production	18 000 barrels/day
Salient development problem	Production only possible when weather permits: reserves difficult to assess
Production facilities	
Platforms	One steel floating
Transport facilities	SPM loading buoy & dedicated tankers (*Theogennitor, Leonidas*)
Service/supply base	Aberdeen
Estimated capital cost	£90 million (to date) £125 million (total)
Approx. operating cost (1982)	£55 million/year

the production facilities via flexible hoses.

The riser system is supported at the surface by constant tension. Riser tensioners tension both the 10 in. central riser column and the individual 4 in. flowline risers. Tensioners were selected to run continuously at about 50% design rating with a 10 ft maximum rig heave when the riser is connected. Individual rotary hoses are used to connect the moving riser head and fixed flanges of the pipework on deck.

Base and subsea manifold

On Argyll there is a riser base acting as a mass anchor and a support for the subsea manifold.

The riser base was designed in two parts – a manifold section containing all automatic hydraulically operated valves, and a permanent base which contains all manual valves. The permanent base attaches to the mass anchor set into the sea bed, and is not retrievable. The manifold section may be retrieved for servicing and replacement of hydraulic valves.

The riser base (mass anchor) is precisely what its name implies. It is a concrete base 40 ft × 40 ft weighing 500 tonnes with a steel frame, vertical guide posts and a central vertical 42 in. diameter conductor pipe. The conductor pipe is used as a support and latching member for the permanent base, and the guide posts assist in locating the permanent base. The sides of the mass anchor slide downwards towards the sea bed and contain saddles for restraining the incoming flowlines.

The permanent base is an extension of the drilling concept permanent guide base, complete with guide wires from its four corner guide posts. The guide base concept has been developed to provide pipework for interconnection of the subsea flowlines with the next element, the riser manifold. Incorporated within the permanent base are manually operated valves for closing off each flowline should the need arise. These are manually operated by divers.

Stabbing into the permanent base is the riser manifold, which is located by the base guide posts. The manifold is latched into position hydraulically. Incoming wells are connected from base to manifold by parallel male to female "choke and kill" connectors and the outgoing export lines connect by hydraulic connections which also serve the manifold. The pipework within the manifold is arranged to connect each well to its individual riser via a hydraulically actuated valve. Each well can also be routed, via a valved tee, into a "ring main" facility. This ring main serves two functions – it enables each well to be produced via any riser, for riser maintenance, and it allows that each riser can be used to service any other riser, for washout purposes. The manifold pipework is terminated on top of the manifold with hydraulically operated connectors. Risers which stab into these connectors are arranged radially around the larger central riser.

Initially, the hydraulic control bundles were connected to the manifold with a hyraulic connector. Problems arose with loosening, due to currents and movement, causing hydraulic leaks and malfunctions. The connections were replaced with individual swivel connectors. The original manifold was assembled with two-bolt clamps with standard API or BX rings. They were repeatedly retorqued because of leaks. Standard API flanges were found to be more satisfactory.

After 3 years of service the original subsea manifold was replaced after a crack developed in the discharge piping which could not be repaired *in situ*. The crack was caused by fatigue from the action of the tensioned riser. The original manifold was designed with a split ring with the intention of keeping the riser tension off the manifold piping.

The studs and nuts holding down the split ring had to be retorqued periodically. Even so, several times the studs broke off flush with the manifold top and had to be drilled and extracted.

The second manifold design eliminated this problem but still needed to be double-nutted to prevent the backing-off of the retainer nuts. A considerable amount of subsea maintenance was carried out on the manifold, as well as a regular schedule of checking and torquing bolts.

The autolocks themselves have also required maintenance. The most frequent activity is the replacement of the seal within the connector usually following latching, after running the riser.

The autolock latches with a pressurized circuit which shifts the latching piston. After latching, this circuit can be depressurized without shifting the latch open. To unlatch, a second circuit is pressurized which shifts the piston back. This circuit is normally left depressurized during production periods. Both of these circuits have redundant circuits to back them up in case of failure. The two circuits enter the autolock through a tee screwed into the body. Once, during a storm, the autolock failed to unlatch so the riser could not be pulled. Unfortunately, the unlatching circuit was broken at the tee while it was depressurized and there was no surface indication of this until the circuit was pressurized to unlatch the autolock. Frequency checks and a design change minimized this problem.

The 10 ¾ in. riser has a universal joint with a short high-pressure hose running through it to allow angular deflection between the riser and the manifold. The hose and the U-joint bushings are replaced annually.

Damage has occasionally been caused to the manifold pipework by loosening of flanged joints, caused by alternating flexible stresses transmitted via the riser during marginal weather conditions. The use of backing lock-nuts has been found to be a simple and effective cure for this problem.

Riser-handling equipment takes the form of a traditional drilling derrick, complete with draw works and accompanying equipment. This is not advantageous to the operation of a floating production unit, because it adds to the instability of the unit during bad weather conditions by raising its centre of gravity. Riser fatigue control is carried out on an annual basis by shipping the riser ashore for a comprehensive non-destructive testing (NDT) inspection. This inspection usually takes the form of a critical visual inspection for obvious defects, a full-body magnetic particle inspection (MPI) of all riser pipes, for detection of surface cracks, X-ray and ultrasonic scanning of all butt welds for cracking of the heat affected zones (HAZ) or weld metal, and ultrasonic wall-thickness checks to detect any erosion damage. In addition to this, the tool joints are specially surveyed for tong damage, thread damage and seal-area damage. Repairs are carried out as necessary. These inspections are closely monitored, often being witnessed by the Department of Energy, Pipeline Engineering Division.

BUCHAN FIELD

The Buchan Field layout is represented on Fig. 2, and the key information summary is given in Table II. The Buchan production riser (as with the Argyll design) is not designed to stay connected in severe weather conditions, but must be recovered and stacked in the derrick. It has been designed to survive 12 m maximum waves with the associated rig offset and current, but recovery of the riser has to be initiated at a considerably lower threshold. However, it is the ability of the tanker to stay loading at the buoy rather than the riser

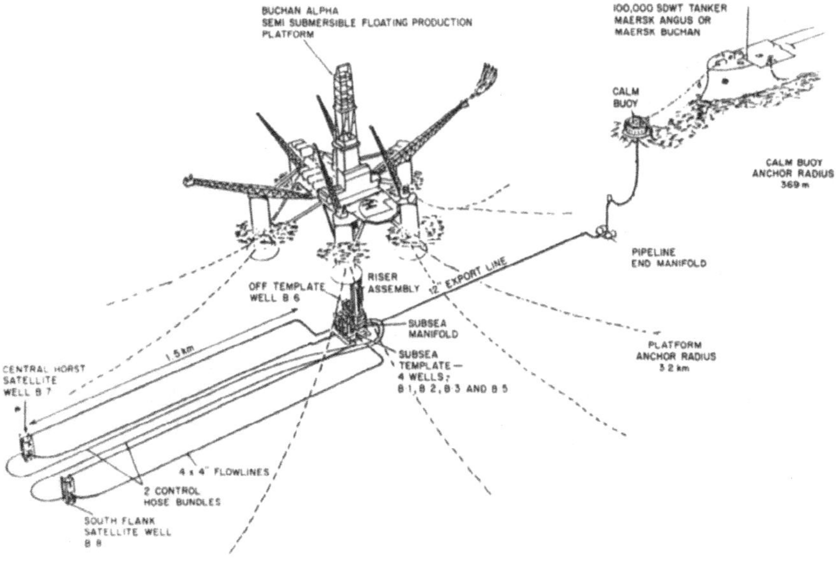

Fig. 2 Buchan Field layout.

Table II
Buchan Field–key information summary

Operators, partners	BP, Transworld, Sulpetro, Clyde, St Joe, CCP, GAOL. Charterhall, Lochiel, Texaco (20 May)
Discovery date	August 1974
Water depth	118 m (390 ft)
Sea-bed nature	Thin varying sand
Reservoir depth	2900 m (9600 ft) to 32000 m (10 000 ft) approx.
Recoverable reserves	
Original	50 million barrels (upgrading probable)
Remaining	48.5 million barrels (31 December) (more with upgrading)
Recovery factor 13% with gas lift	API gravity 33.5° GOR 300 scf/barrel Sulphur content 0.8%
Production start	May 1981
Peak production	47 000 barrels/day in 1982 (72 000 barrels/day in summer months): design max. was 72 000 barrels/day
1981 production	40 000 barrels/day
Saliant development problem	Difficulties of conversion of existing semi-sub for a complex field
Production facilities	
Platform	One steel floating
Transport facilities	Catenary Anchor Leg Mooring Buoy dedicated tankers (*Maersk Angus*, *Maersk Buchan*)
Service/supply base	Dundee
Estimated capital cost	£225 million (to date) £250 million (with gas lift)
Approx. operating cost (1982)	£27 million/year

which is the limiting factor in the production from the field. Before the introduction of gas lift the riser bundle consists of a central 12 in. export riser and ten 4 in. risers. The ten 4 in. risers consist of one production riser from each of eight wells and two service risers.

Each riser is independent of the others and can be run and tensioned separately. However, the ten 4 in. risers are constrained to deflect the 12 in. riser by being passed through guide funnels. These guides are supported on arms mounted every 7.6 m along the length of the export riser. Hence, although each riser in the bundle is independent in the axial direction horizontally, the bundle tends to deflect as a single composite member.

To reduce bending stresses at the base of the risers, a universal joint is mounted in the export riser directly above the manifold. In addition, the bottom 15 m joint on each production riser is made from extra heavy tubular. With the introduction of gas lift, facilities exist to clamp a 2 in. gas lift riser to each production riser, and the guide funnels are sized to accommodate the pair.

The riser concept was designed for operations flexibility. Only sufficient risers need to be run to accommodate flow and, if required, individual risers can be pulled and replaced without having to recover the entire bundle. On the other hand, riser handling time during the running or recovering operation is increased by the need to handle individual lines sequentially.

The riser was the subject of an immense amount of design analysis, and model tests were conducted on the bundle to establish drag and inertial coefficients on individual risers. These data were then used, in the stress analysis programme, to predict the bending behaviour of individual risers as well as the gross behaviour of the bundle. A spectral fatigue analysis was conducted to determine fatigue life, and this was supported by fatigue testing of a threaded riser connector. In addition, during operation, riser behaviour is monitored by instrumentation measuring parameters such as the deflection of the universal joint and the pipe wall strain at two locations on the riser.

Buchan template and manifold

To meet the production drilling programme it was essential to build and install the drilling template for the cluster of subsea wells during the summer of 1977.

The template design was therefore simple for ease of fabrication. Since the final layout of the wells had not been determined, the template had to provide a number of drilling slots with the facility to support flowlines. Design, fabrication and installation of the template took a total of 4 months.

An eight-slot template with a 5.2 m space between well centres was chosen with the underwater manifold occupying one slot. Provision was therefore made for five template wells and two spare slots in case drilling was abandoned on any location. The template is located on the sea bed on a 0.65 m wide skirt around the base of each 1.22 m diameter conductor guide. Four 12 in. piles are driven to secure the template. To level the template, provision was made to jack up on its piles, in the event of a sloping or uneven sea bed.

For the installation the 90 tonne template was transported to the field on a flat-top barge. At the location the template was lowered into the sea by a crawler crane secured on the deck of a large supply vessel. It was then winched down to the sea bed, where divers confirmed the orientation and position, approximately 7 m from an existing exploration well which was to be tied in as a producer. The piles holding the template in position were driven using a small underwater vibro-hammer. No levelling was required, and the template was secured to the piles by the inflation of annular packers. Overall, the installation time was about 4 days.

Drilling and construction work validated the basically simple concept and, if the template were redesigned, very few modifications would be introduced.

Following the template installation in August 1977, drilling commenced on the four template wells. This activity extended over a further 14 months, and also included setting the 90 m of 30 in. conductor to act as a foundation for the underwater manifold. The wells were temporarily suspended pending delivery of the subsea trees.

Simultaneously two satellite wells were drilled by a second rig during 4 months in the summer of 1978. With the modification of the original programme, of five template wells plus one southern satellite, to four template wells, plus two satellites (one of which was drilled back into the central formation) the drilling schedule was maintained.

Dual-bore split trees were used. Each tree weighs 52 tonnes and was installed as a single unit. The lower tree contains a dual manual master valve block; the upper carries a six-valve block – two master, two wing and two swab – all hydraulically operated and fitted with external indicators. A bypass loop carrying a further three valves is mounted on the lower tree and provides the link between the dual 4 in. flowlines back to the manifold and the upper tree.

Before introducing gas lift, the bypass valve can be left open to allow both flowlines to carry oil and to permit pigging to remove wax deposits.

In the event of major equipment failures, the upper tree can be recovered leaving the lower half in place. This task would require diver intervention to disconnect the control circuits of the flow loop flange.

HUTTON OILFIELD

The tension leg platform (TLP) paves the way for oil production in water depths beyond the feasibility of conventional fixed structures

(Fig. 3). It neither floats entirely free nor stands entirely rigid. It is the world's first commercial application of a new concept in offshore technology.

The production platform has to be capable of withstanding winds of up to 110 miles* per hour and waves of up to 105 ft high. It weighs almost 50 000 tonnes and its steel deck structure outflanks a football pitch. After 5 years of research, Conoco concluded that the concept was feasible and could be designed to be reliable – equally significantly, the cost of a TLP should be relatively insensitive to water depth.

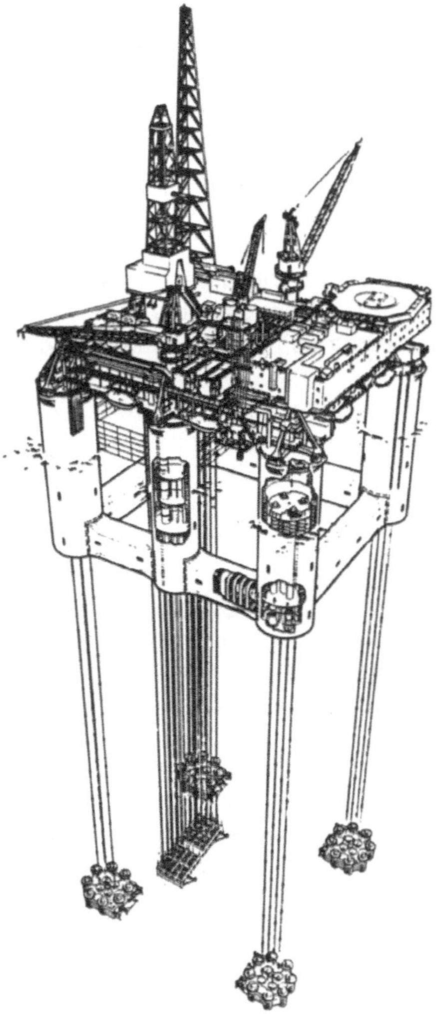

Fig. 3.

*1 mile ≈ 1.6 km.

The system allows most of the assembly to be accomplished onshore or in sheltered, near-shore waters before tow-out to the production site. This offers obvious efficiencies over on-site assembly in rough waters far from shore.

The hull had to have the characteristics of a normal semi-submersible drilling rig, and yet be capable of remaining in the extreme conditions of the North Sea for 20 years without once returning to dry dock. The design of the steel tension legs was entirely new. A plug-and-socket mechanism had to be designed to allow each leg to be locked into position and held under tension. Yet each leg also had to be capable of release for removal and inspection. Elastomeric technology first developed for the space programmes had to be called up to form the special rubber-and-steel flex joints which would allow the unique mooring system to absorb the natural movement of the waters.

The TLP consists of four basic elements: the deck, the hull, the mooring system and the well template.

Deck structure. A single integrated dock with a mezzanine carrying all the process and utility equipment plus external modules for the drilling rig, power generations, accommodation, helideck and flare structures.

Hull structure. A six-column semi-submersible structure which provides buoyancy and contains a ballasting system to maintain platform trim.

Mooring system. Sixteen vertical tubular steel tethers, 260 mm in diameter, running from the mooring compartments in the hull to piles and grouted foundation templates on the sea floor. The mooring system is designed to enable the legs to be removed individually for inspection and maintenance without operations being affected.

Well template. A 32–well-slot template was installed in May 1981 to enable production wells to be drilled before the installation of the TLP.

For final connection five tugs took the Hutton TLP on its final journey to the field. There to meet it were two large semi-submersible crane vessels, the *Hermod* and the *Balder*, whose task would be to hold the TLP in position while the tension legs were lowered from the corner columns and linked to the foundation templates deep below on the sea bed.

With the platform ballasted low in the water, the first tension legs were latched into position, guided on their way by engineers using acoustic positioning equipment and underwater cameras. Then, with one "leg" connected to the mooring system from each corner of the structure, the Hutton TLP was gradually brought into tension by releasing the ballast. It was 7.57 a.m. on Friday, 13 July 1984 when vertical heave was suppressed. Within 2 days the remaining 12 tension legs had been latched and tensioned. Attention could now turn to preparation for first oil production.

As the platform equipment was already fully operational, remarkably little remained to be done. A connection was made to the pre-laid pipeline which was to take Hutton oil via the North West Hutton platform into the Brent pipeline system and on to the Sullom Voe terminal in the Shetlands.

Next the drill rig was carefully positioned over the first well – one of ten already drilled from a floating rig through the sea-bed drilling template. The well was connected to the platform, the steel casing was perforated, and the Hutton TLP was ready to produce its first oil.

PIPELINE/FLOWLINE LAYING METHODS, GENERAL REVIEW

Lay-barges generally provide the fastest, most economical means of installing pipelines. For instance, with single-line installations, speeds of 1 mile/day are possible.

For applications where small numbers of lines must be brought to a structure, the lay-barge is ideal. Pipe sizes of 6 in. or more are commonly placed in this manner. While the lay-barge is well suited for the placement of a few large lines, it is not recommended for use with multiple small-diameter flowlines. Each line to be placed requires one pass of the barge over the pipeline route. The repeated mooring operations dictated by this method extend the laying time and cost to unacceptably high levels.

A further constraint on this type of installation is mooring of the barge in areas of multiple flowline concentration. As each additional flowline is laid, the risk of rupture or snagging by a mooring line is increased.

The lay-barge is best suited for installation of the 12 or 14 in. diameter export line. Laying of the pipeline should be done, if possible, before any other flowlines are placed, so as to avoid mooring complications. This method has considerable merit for applications where multiple lines are to be laid to a subsea structure. Its main advantage is the relatively short offshore work time compared with individual line-laying methods. Other advantages include the capability to weld, inspect and fully pressure-test onshore, no mooring of work vessels during installation, and the additional protection provided for the flowlines by the carrier pipe.

Submerged tow requires encasement of the flowlines in a carrier pipe for buoyancy and protection during launch and towing operations. Each end of this carrier is fitted with a tow head, which provides a tow point and pressure seal for the flowline bundle. The assembled package is pressurized with nitrogen to the site sea-bed pressure. This pressure is periodically monitored during the tow to check the watertight integrity of the carrier. During launch and tow operations the flowline bundle is negatively buoyant, 1.5 lb/ft*. To ensure this, strict weight control is exercised throughout the assembly operation.

Fine-tuning of the bundle weight, and in turn its buoyancy, is accomplished using chains attached to the carrier pipe. Once launched, the bundle is allowed to settle to the sea floor. As the attached chains touch bottom, their weight is supported by the sea floor and the pipe settles above bottom. Once an equilibrium point is achieved, divers are used to remove excess chain to obtain a final weight trim before towing begins.

Upon completion of the weight trim, tow to location begins. Tow depth is controlled by tow speed, 3.5–5 knots and the hold-back force applied by a trailing tow vessel plus the length of towline paid out. The major disadvantages of this method are the additional cost of carrier pipe and towing equipment, together with the limitations in length for towing and the limitations in length of yards offering this kind of technique.

Although the method is relatively new, it has had substantial field exposure. Occidental and Shell have had good results with submerged tows on Claymore and on the Underwater Manifold Centre (UMC). Occidental has two multiple bundles laid at Claymore, and Shell are in the final stages of placement of three multiline bundles to their UMC. Hamilton Brothers recently lost a flowline that they were attempting to install by this method. Later evidence appears to indicate that the spiral-weld carrier pipe used was at fault, rather than the method.

The cost of carrier pipe and tow equipment does not appear to rule out submerged tow on a cost basis, because these are offset by the much lower cost of onshore welding and the short duration of mobilization of smaller, less-expensive offshore boats and equipment. The preliminary cost data given in the flowline appendix

*1 lb ≈ 0.45 kg.

material shows the method to be competitive.

The significant offshore time saving possible with this method is a major factor in its favour. It is difficult to judge accurately the total impact this could have on installation, other than to point out the reduced effect of weather downtime of the method compared with individual line-laying methods. Based on existing cost data and the operational advantages outlined above, it is recommended that this method be used for placement of satellite reservoir flowline bundles.

Reel-ship installation is judged to be the most suitable method of installation of satellite flowlines. Use of either flexible or steel flowlines has been examined. From a cost analysis, flexible flowlines are preferred.

From an installation standpoint the dynamically positioned reel-ship approach is ideal for this portion of the development. A dynamically positioned reel-ship avoids mooring complications and can provide extreme accuracy in flowline placement. Additionally, the lay time for single and dual flowlines is very low. Speeds of up to 1000 ft/h are possible in good weather.

As with conventional pipe-laying operations, the cost can vary dramatically depending on existing market conditions. An in-depth analysis of conventional steel flowline and flexible flowline lay cost is recommended before actual installation.

Flowline connection methods – diver versus remote

Diver-made connections are relatively common in the North Sea. Both Argyll and Buchan made extensive use of divers in their connection operations. Many other North Sea operators have shown themselves to favour diver operations, water depth permitting, and Balmoral and Highlander will soon be other examples.

Diverless connections have been developed because of the desire to reduce the costs of diver intervention, and also because of the need for a tie-in method that can still be used at depths where diver intervention is either impossible or impractical. The most recent application of this technology is its use on Shell's UMC. Diverless connections have also been used on the Ninian and Brent Fields (see Fig. 4). From a functional standpoint a remote pull-in tool has limited tolerance for error. Use of this approach also prohibits individual repair or replacement of flowlines due to end-connections. Another major drawback to pull-in tools is the required presence of a drilling vessel to operate the tool. If this vessel is already on location, then no major additional cost is incurred. Generally, however, this is not the case, and availability and mobilization costs of suitable vessels for

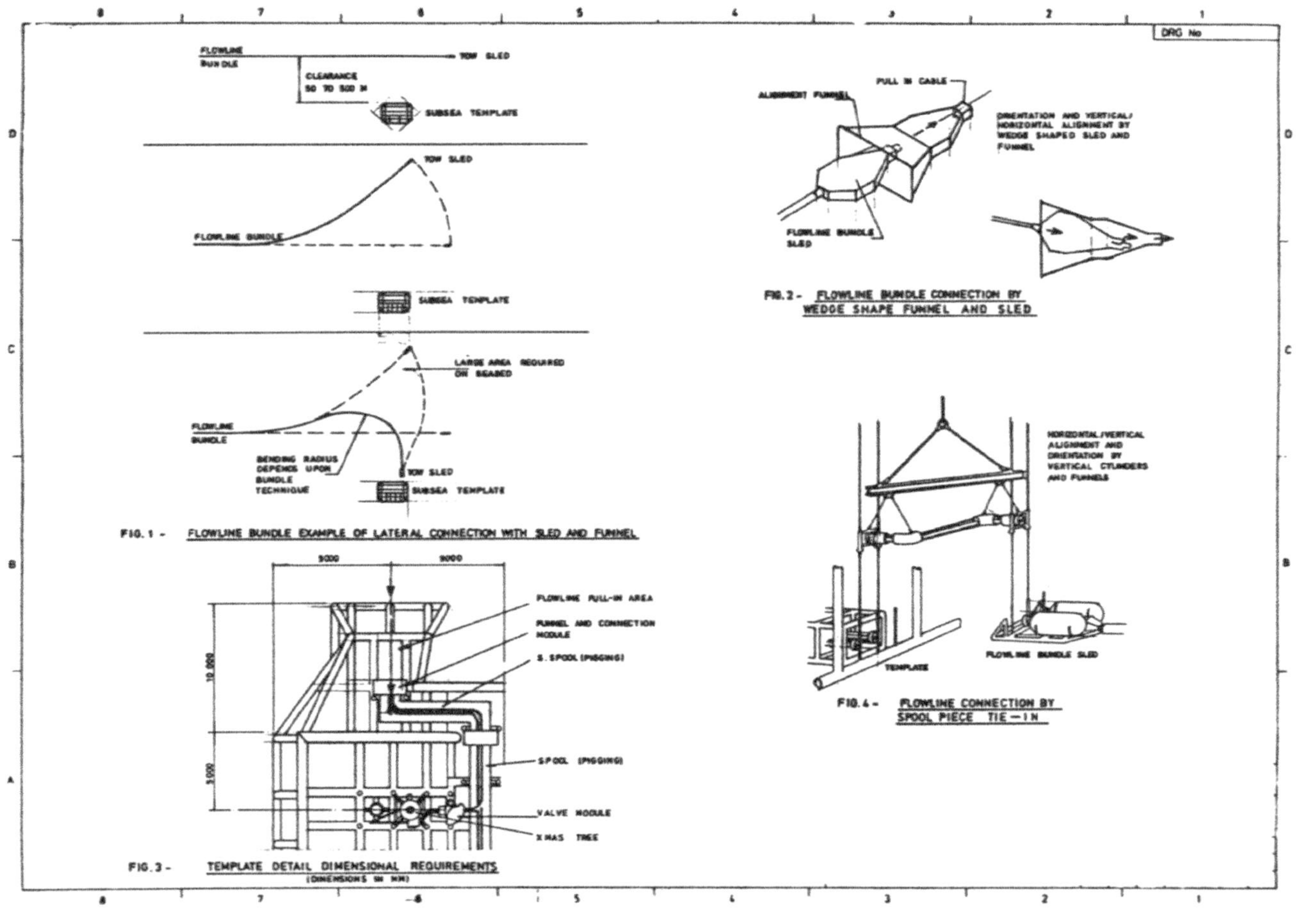

Fig. 4 Flowlines connection method.

the remote pull-in tend to favour the use of alternative methods using divers.

One of the major inconveniences of remote tie-in methods is that they require template space. For a small field with seven or eight satellite wells to be connected, the space taken will increase the template by 30–40%, with an obvious direct and similar impact on cost (see Fig. 4).

To overcome the problems associated with offsets and misalignments between flowlines and template interface points, the use of misaligning flanges on all lines is recommended. To protect the template and minimize damage to flowlines in the event of dragged-anchor incidents, pressure-balanced safety joints are recommended for all lines. As an alternative, smaller lines, of less than 6 in. diameter, could be fitted with machined bolt connections to provide a safety break point rather than using a pressure-balanced joint. These joints, called load-limiting connectors, separate when a preset stress is reached, and safety valves on either side of the joint automatically close to prevent pollution damage.

For connection between the end of the flowlines and the piping in the template structure, many engineers will recommend the use of rigid, made-to-measure, spool pieces. An example of installation is shown in Fig. 4. When the soil conditions are adequate this is a good technical solution, but it is diver-intensive for two main reasons:

- the need to measure, by diver, the final dimensions of the spool (doubling the diver time);
- the need to do a final weld on the spool, which requires more time, more equipment and more expertise offshore.

In addition, the installation of large-diameter spools and long spools, as shown in Fig. 4, necessitates complex and expensive handling techniques.

A very attractive solution is to use flexible spools. These have been used extensively in the North Sea, and were used in 1983 by Texaco and Sunoil in fairly large quantities. In addition, when soil conditions are poor and unstable, flexible spools are the only proper answer.

Trenching and burying

Large-diameter lines and flowlines in deep water are often left untrenched in the North Sea, and this will also be a tendency for the future. There are, however, many cases where trenching will still be required.

Fishing activity in the North Sea is a significant factor in determining the trenching requirements, and the fishing industry is not static; developments over recent years have had an important influence on pipeline design. Bottom trawling is the type of fishing which most affects a pipeline, but traditional otter-board trawls have been demonstrated to be relatively harmless to pipelines more than 16 in. in diameter. As time goes by, Dutch Beam trawls are being used in more northerly latitudes, and forces on the pipeline resulting from this activity must be considered during design. Vessels travel at up to 7 knots when trawling, and the combined weight of beams, skids, chains and nets may be 11 tonnes. Clearly, the potential for damage to an exposed pipeline is considerable.

In more-northerly altitudes, especially on the American continental shelf, designing for iceberg scouring must be considered. In these areas deep trenching may be required to give a reasonable degree of confidence that damage will not occur. In shallow waters, especially in exposed areas with long reach, pipeline stability requirements or soil movements will often dictate trenching. Various aspects of navigational activity must also be a design consideration. If ships of deep draught operate in relatively shallow water, then pipelines will generally need trenching below the sea bed. There is little that can protect a line from a heavy marine anchor, but any likelihood of vessels anchoring in the vicinity will generally dictate trenching. This is especially the case if wire-moored vessels are anticipated, because trenching will considerably reduce the risk of damage by wire-cutting.

Spans will occur when an untrenched pipeline crosses an undulating sea bed. These may be corrected with sandbags and mattresses, but there is always the danger of scour around discrete supports. Rock dumping overcomes this problem, but may cause high stresses in the pipe. The most satisfactory solution will often be found in post-trenching of high spots.

Trenching may be performed either before or after pipe-laying. Normally pre-dredging is limited to shore approaches, but trenches have been pre-ploughed in deeper water. However, the major thrust of current development work is in post-trenching techniques. These may be divided into three basic categories: the jet sled, the plough and the mechanical trenchers.

SUBMERGED PRODUCTION SYSTEMS

Conceptual design review

Zakum Subsea

The Zakum Subsea Production Scheme commenced with the spudding-in of the well in August 1969 and continued until April 1972, when the scheme was decommissioned, leaving the subsea wellhead flowing direct to an adjacent platform. During this period a variety of subsea production equipment, operations and services were evaluated (see Fig. 5). The objectives of the scheme were:

- to achieve production from the well using subsea methods;
- to gain experience of subsea production methods and equipment, so as to evaluate their practicability and relevance to future operations.

The range of equipment and techniques investigated covered most aspects of subsea production. In addition to primary production equipment (e.g. wellhead, valves, actuators and flowlines), a range of ancillary equipment and support services (e.g. separators, electric power, instrumentation and diving support vessels) were also included. The full range of equipment and services consisted of:

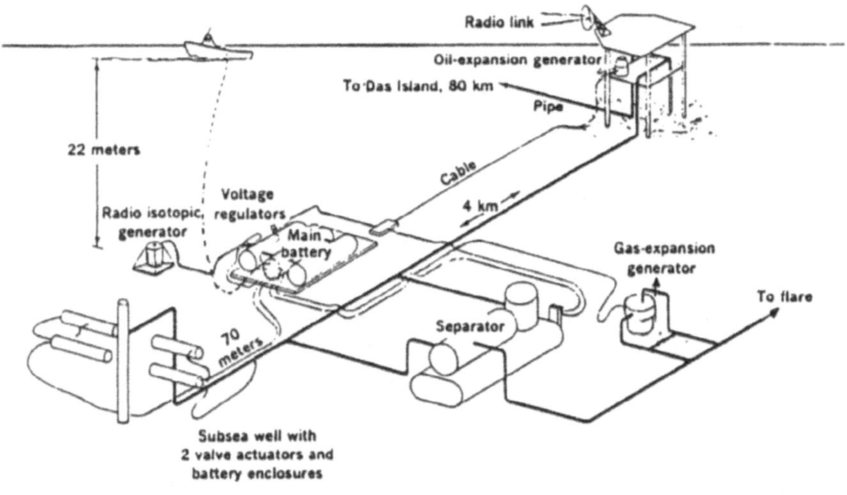

Fig. 5 Schematic diagram of Zakum subsea production scheme.

wellhead production;
oil/gas separators;
gas-disposal systems;
valve actuators;
instrumentation and communications equipment;
electric power sources and distribution;
piping, manifolding and ancillaries;
wireline operations;
diving services;
surface support vessel.

The offshore conditions for operating a subsea production system at Zakum were not difficult, and the water was only 20 m. These conditions permitted the use of an air diving service for support on installation/operation and maintenance. Production conditions were also straightforward with an established shore base (Das Island) near to the centre of operations. However, the scheme provided a considerable amount of experience which has been applied to subsea operations in deeper water and under more-severe conditions. During the period August 1969 to April 1972 the major work effort was as follows.

Period	*Major work effort*
August-October 1969	Well drilled and completed using a subsea wellhead
January-March 1970	First series of wireline trails
November 1969–August 1970	Fabrication work and installation of main equipment
August-December 1970	Commissioning
December 1970–December 1971	Operation and evaluation
January-February 1971	Second series of wireline trails
March-April 1972	Decommissioning

Exxon SPS and Shell/Esso UMC

Shell and Esso's basic project philosophy was to use proven technology as much as possible in developing a design with emphasis on safety, reliability, flexibility and ease of maintenance. However, not all existing components met all the North Sea project criteria for flexibility and reliability sufficiently. In-depth design studies and intensive post-construction testing to satisfy these requirements have occupied the two companies since 1974. Indeed, the start of work on the UMC concept goes back even further. In 1969 Exxon began a development programme which culminated, in 1974, with a pilot test of its Submerged Production System (SPS) in 170 ft of water in the Gulf of Mexico. The success of this US$81 million project included remote flowline connection and robot maintenance. With 18 months of production from three wells this was a major factor in the development of the UMC.

Subsea experience. Shell's input to the UMC project was based on long experience of subsea-tree operation offshore Brunei, and a pilot test of a 1 atm (101 kPa) encapsulated manifold.

The UMC is similar in appearance to Exxon's SPS, but offers additional features – in particular, the ability to serve wells drilled directly through the template and/or remote satellite wells. Essentially the UMC provides four basic functions:

- drilling of wells through the base structure and tie-in satellites;
- collection of fluids from producing wells for delivery to the South Cormorant platform 4 ½ miles away;
- distribution or treated sea water from the platform to selected injection wells;
- maintenance of valves and controls without the use of divers.

The basis for the design was a nine-well manifold with each well bay available for template or satellite wells. These wells could be either producers or water injectors. The structure weighs 2200 tonnes and measured 52 × 42 × 15 m. Production capability is 50 000 barrels/day of oil, and the water-injection capability is 56 000 barrels/day. The water depth limit is currently 1000 ft, but it can be readily adapted to suit 4000 ft water depths (see Figs 6–8).

The base structure is a tubular steel frame which supports the other subsystems and assures alignment for connection (it has 120 funnels for running guide posts). It also operates as a drilling and casing guide frame. The structure has nine well bays, four on one side and five on the other, separated by a central track on which the Remote Maintenance System (RMS) runs. Installing choke and "S" spool pieces in any well bay enables satellite tie-ins.

Well completions. On Central Cormorant all of the well completions have two 3 in. internal diameter tubing strings, each with a retrievable subsurface safety valve. As a back-up, a pumpdown insert safety valve is also provided. Other features of the downhole completion are upper and lower members for circulating fluids. There is also a dual detachable packer so that the lower assembly may be left undisturbed during service work. This set-up permits either dual- or single-string flow, and allows wells to be used as producers or injectors. All of the completions will be geared for TFL (through flowline) or pumpdown tool intervention. In addition to the back-up safety valves, TFL will be capable of changing valves, setting and

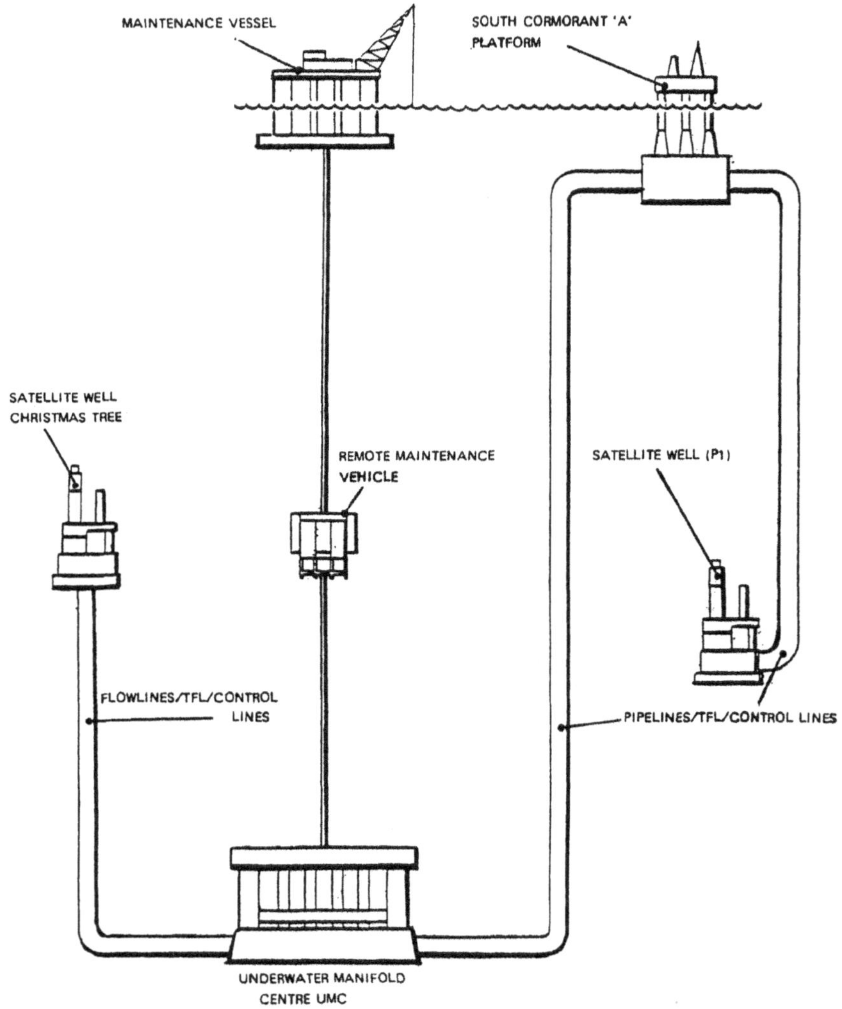

Fig. 6 UMC – schematic view of underwater production system.

retrieving plugs, cutting wax and running downhole surveys.

By using TFL techniques the need for vertical access for servicing operations, which would be expensive and weather dependent, is eliminated from a semi-submersible rig. Before going offshore all downhole components have been extensively tested for reliability and performance.

Christmas trees were also subjected to extensive interface testing. The trees feature dual 3 in. strings with dual master valves and a TFL loop. All seals are metal-to-metal and connections are remote diverless. Working pressure is 5000 psi*.

*1 psi ≈ 6.9 kPa.

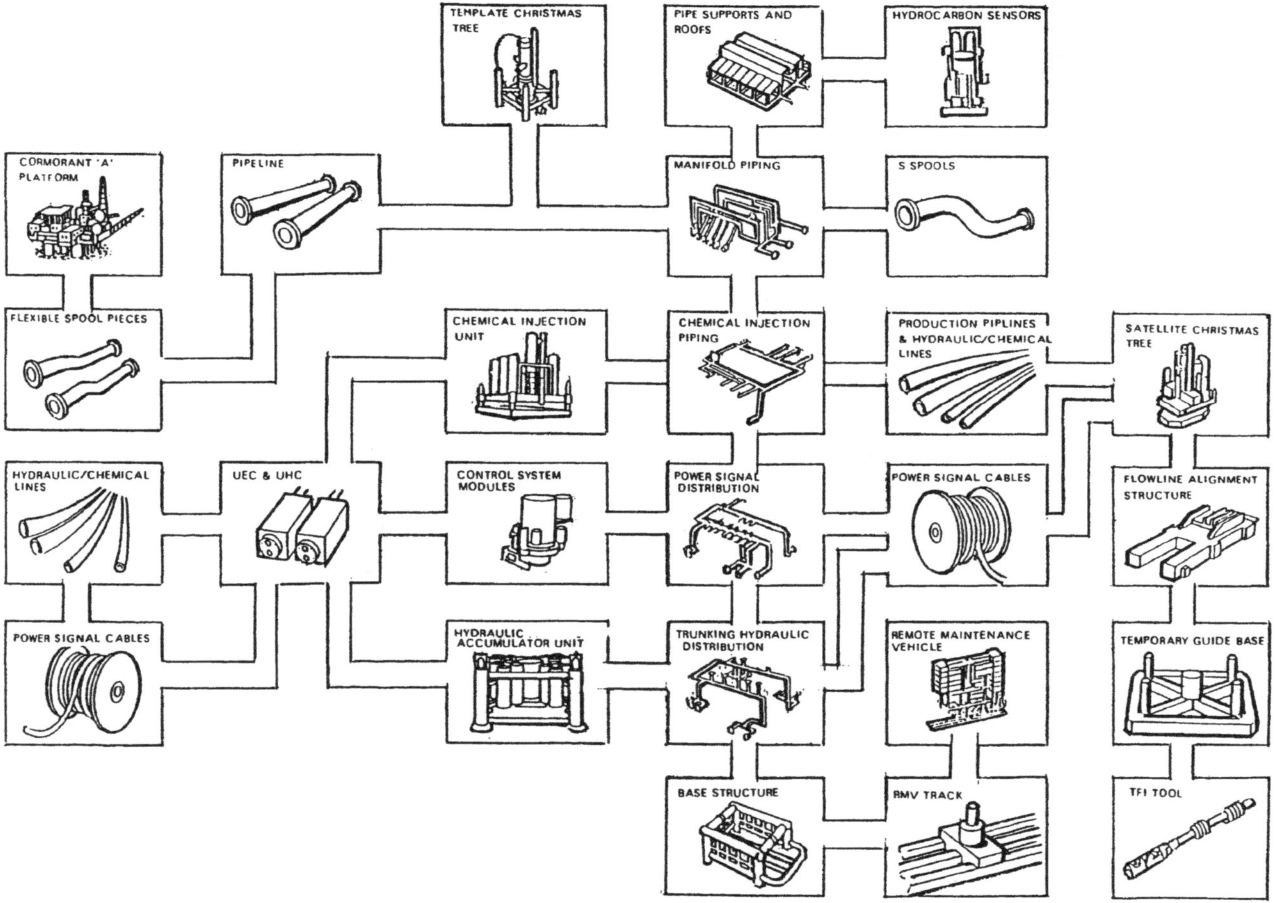

Fig. 7 How the UMC fits together.

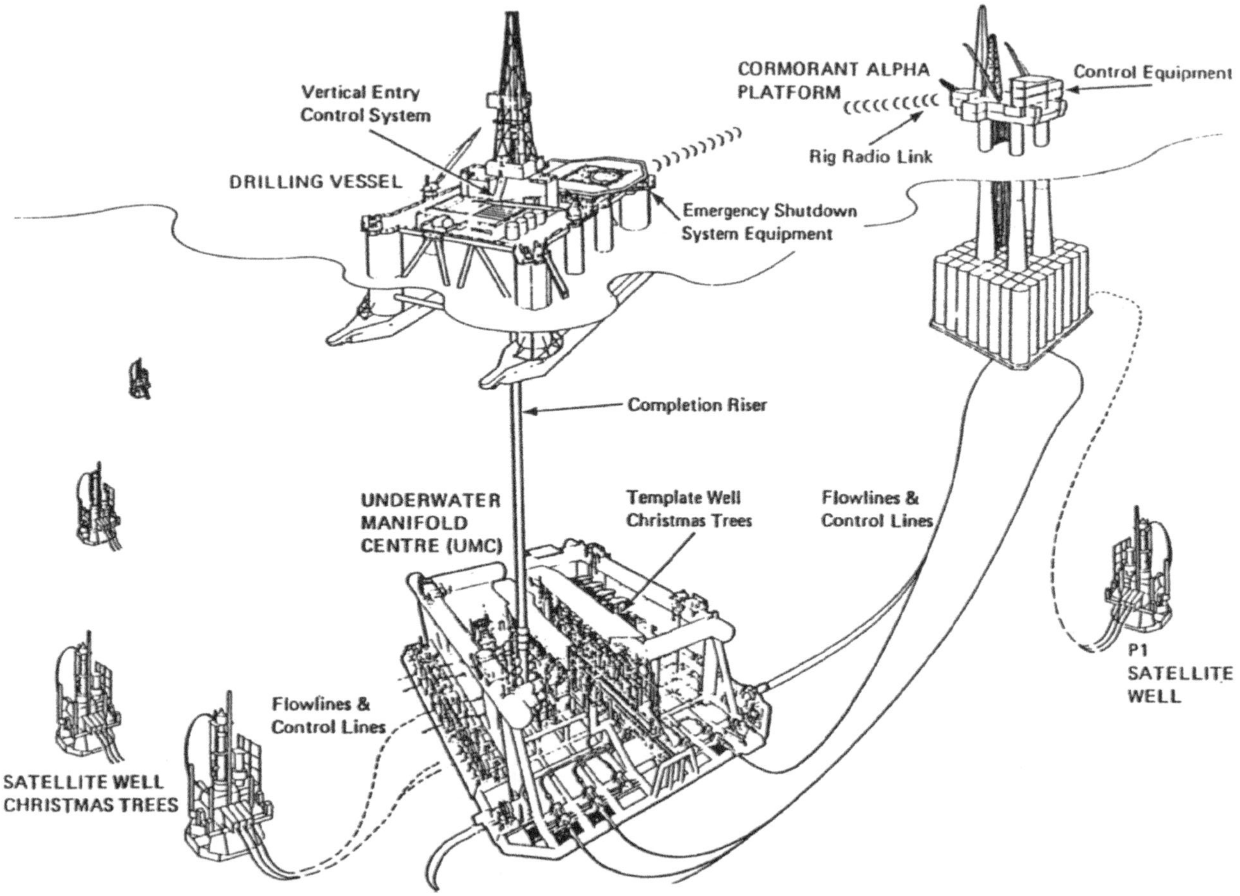

Fig. 8 Rig location during template well completion operation.

Each well bay on the UMC has a removable cover to protect the Christmas tree, particularly during simultaneous drilling and production operations. The flexibility of the system is such that it is possible to replace a well drilled through the template with a satellite well.

Production manifold. The production manifold on the UMC is situated on either side of the central maintenance track. On the manifold there are an 8 in. oil header for collecting oil production, an 8 in. water injection header and two 3 in. headers for well testing and TFL access. In all, there are 110 valves and diverters.

Roofs designed to trap any hydrocarbon leaks cover the manifold and provide protection. In the event of a leak, sensors in the roof sound an alarm in the platform control room. Any hydrocarbons that accumulate as a result of a leak can be recovered back into the 3 in. header system, using valving and blow-down procedures.

Packed valve assemblies serving individual well bays and maintainable by the manipulator were used in the SPS project, but in a different configuration.

Two methanol-injection packages are installed on diagonally opposite corners of the structure, with hydraulic accumulator packages for the control system at the other corners. This arrangement ensures 100% redundancy, particularly in the event of damage from falling debris. However, it would be unlikely that debris would hit both methanol and hydraulic accumulator packages.

The UMC is connected to the South Cormorant platform 4 ½ miles away via two 8 in. production lines, one 8 in. water line and two 3 in. lines. There is also a hydraulic/methanol hose bundle together with two electric power and signal cables. Dual 3 in. flowlines serve the satellite completions. During design of the pipeline system, consideration was given to the effects of cold crude (and line stability during trawl-gear impacts). Wax and hydrate problems could result because of the distance to the platform, particularly for the satellite wells.

Electric couplings. Inductive couplings have been used because they do not require any mechanical link to transfer the power or signal energy. Each half of the coupling houses what is, essentially, half a transformer. This approach was adopted since it was considered to be more reliable than pin-to-pin connectors.

North-East Frigg/gas production

Elf's development in the Norwegian North Sea is located just on the border between Petronord-Statoil's 25/1 block and Esso's 30/10 block. The gas field, with recoverable reserves of 8000 million m^3 has a planned 5-year production life at 6 million m^3/day. This field development is the first subsea development of a gas field and the first use of articulated tower concept as a service platform.

Development plans. The design called for six subsea wells installed on a template, to be controlled from an unmanned articulated tower located about 150 m away. Production is remotely controlled via radio link from the (TCP2) platform on Frigg.

The close proximity of the giant Frigg Field's gas accumulation, to compensate for possible shutdowns on North-East Frigg, was an incentive for the unconventional development scheme. In particular, the scheme would be more suitable to the small size of the reservoir. The approach resulted in a subsea production system consisting of a template to support and protect six wells, a manifold and a 16 in. gas line which links the manifold to the Frigg treatment platform (TCP2). An articulated field control station (FCS) supports the main function of controlling the subsea equipment and treatment. Metering and boosting modules are installed on the TCP2 platform to process the gas (see Fig. 9).

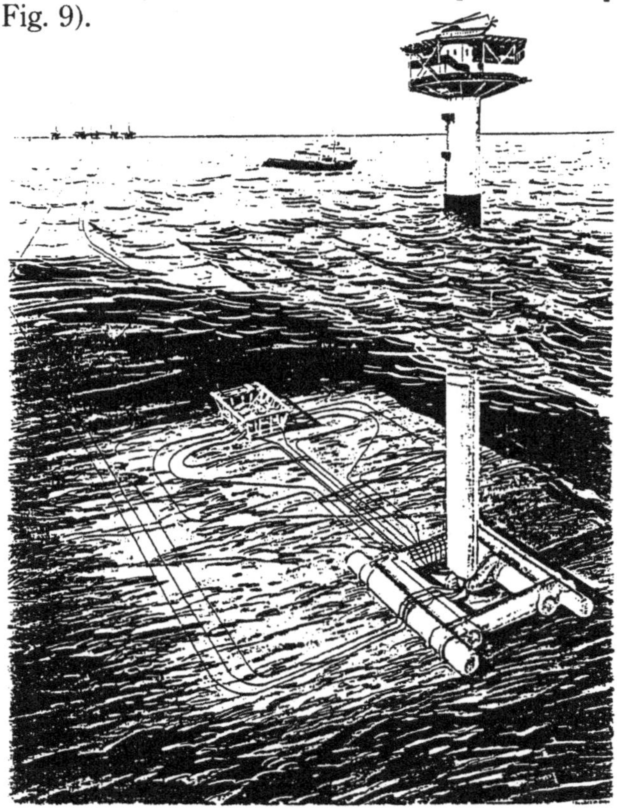

Fig. 9 The North-East Frigg Field general layout.

Apart from processing 6 million m^3/day of gas, the treatment modules on the TCP2 platform also have a daily throughput of 40 million m^3 of condensate and 18 million m^3 of water/methanol. The gas is then transported to the UK through the submarine pipeline system via St Fergus.

As the reservoir characteristics were well known, the wells were not individually tested and no control chokes have been installed. Of the six wells, five are directional and one is vertical. The wellheads are conventional 13 ⅜ in. 5000 psi working pressure wellhead housings and a 9 ⅜ in. casing hanger. The Christmas tree is a wet monoblock unit including the wing valve. The template measures 17 × 30 × 8 m and weighs 350 tonnes. It accommodates six wells plus two spare slots (see Fig. 10).

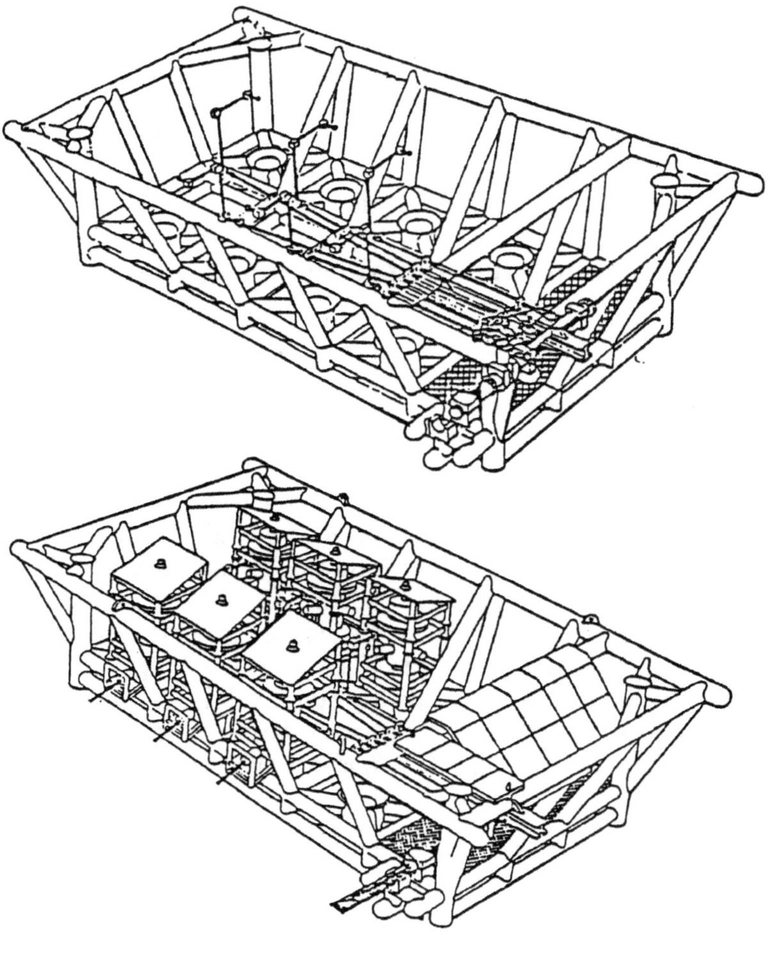

Fig. 10 North-East Frigg template manifold with and without Christmas trees installed.

The control system is designed for control of subsea equipment during normal production operations, such as routine maintenance and special operations and testing of subsea valves. It is automatically controlled and incorporates automatic safety loops.

Articulated tower. The articulated tower has a concrete gravity base (45 × 42 × 6.5 m high) which supports the universal joint. It has a dry weight of 6800 tonnes. As the soil conditions are good, no specific anchoring or grouting is necessary. The universal joint can be connected or disconnected from the base of the tower by a remotely controlled hydraulic locking system. An 8 m diameter cylindrical steel column is the largest component in the tower. In the lower sections of the columns are heavy ballast material, and buoyancy and stability tanks. Water systems are housed inside the column. The column also supports the risers for the umbilicals, a fendering system at sea level and the field control system on top. The station consists of four decks: a methanol deck with methanol pumps and storage tanks, a main deck with the well control manifold, a living-quarters deck and a helideck. Within the remote-control station are the hydraulics for the subsea control, monitoring and communication systems between the main Frigg Field and the control station.

Installation methods: buoyancy, positioning, downhaul

Shell/Esso UMC

The UMC design followed the Exxon SPS prototype concept for transport and launch by using a barge. After launching, the UMC is ballasted and lowered to the sea bed under controlled negative-buoyancy conditions. It was originally planned to allow the UMC to sink and swing under the drilling vessel. It would then be lowered to the bottom using the vessel's draw-works and a large-diameter synthetic line, as for the SPS. Due to the large mass, and added-mass effect, of the UMC, a mechanism is needed to prevent a shock load on the lowering gear. The simplest and most reliable solution for dealing with the relatively light hook loads of the UMC in water (often 50–200 tonnes) and very large inertial mass effect (2000 – 8000 tonnes and difficult to quantify) was to use a very elastic nylon sling system.

Evaluation of alternative installation methods. During the detailed evaluation of the UMC installation procedures it was noted that several factors differed significantly from the Exxon SPS case. First, seastate conditions in the northern North Sea were very severe and often unpredictable. Therefore, the rather complicated and lengthy rigging and swing under (keelhauling) procedures would leave the UMC vulnerable. It could be a long period after launch before lowering the UMC to the sea bed. A second major consideration was the possible interference with the semi-submersible drilling vessel's anchor lines, and the vessel's stability during the swing-under procedures.

However, by the late 1970s a number of very large capacity crane vessels were becoming available, and these provided a very acceptable alternative. Several variations of crane vessel installation were compared, along with the drill vessel/swing-under method.

Launch and lower with a crane barge (see Fig. 11) was selected as the most appropriate method for the Central Cormorant UMC. This method had the advantage of handling the UMC as-built with no structural modifications. It did not require expensive rigging, and it was not sensitive to the added mass when a suitable soft-spring nylon rigging system was used. The offshore installation could be carried out by a number of contractors. There would be no need for

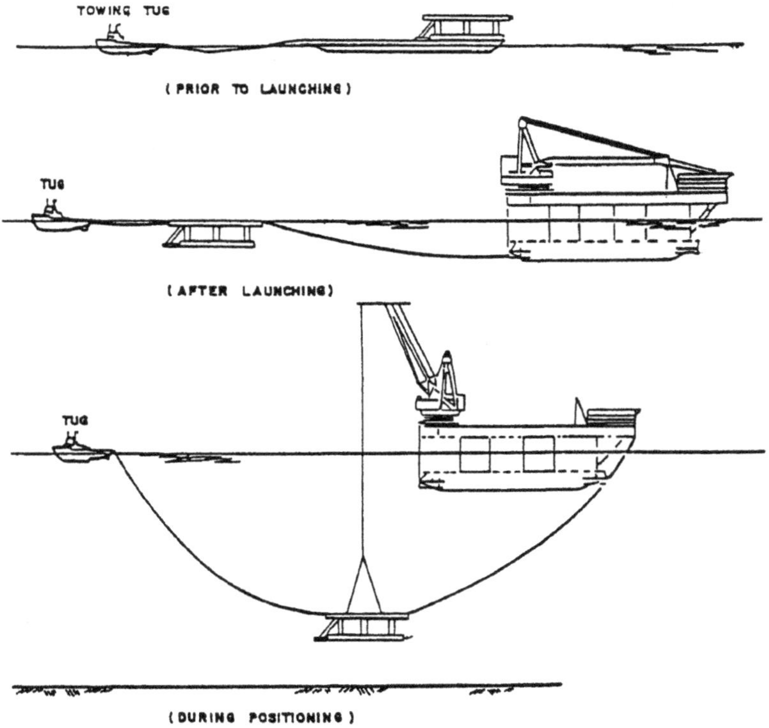

Fig. 11 UMC installation. Launch from barge and lower by crane-barge.

fendering or diver intervention in the wave zone if proper rigging and platforms were used. The entire installation had a number of comparable examples in the North Sea concerning the launch, positioning slinging and pulling of the structure with the derrick barge to centre it under the crane hook.

Installation of the Shell/Esso UMC. Once the installation method was confirmed, a contract was let, and the programme was divided into two parts. Phase I consisted of engineering design and procedures. Phase II consisted of the final detailed engineering and procedures, provision of equipment/vessels, onshore preparation and offshore implementation.

The primary considerations in developing the installation plans and procedures included the following.

(1) *Seastate limitations.* The limitations determined for tow, launch, rigging and lowering. The timing of rigging/surface handling between launch and actual lowering was of greatest concern (overall 1.5 m significant wave height).
(2) *Vessels and equipment.* A 300 × 90 × 20 ft cargo barge with launch ways and haul-off rigging was selected. This was considered the most appropriate to transport and launch the 2200 tonne UMC. Three derrick vessels were considered – the 3000 ton* ship-shape *Odin* and the 3000 ton semi-submersible crane vessels (SSCV) *Hermod* and *Balder.* The SSCV type vessel was chosen because it is less sensitive to seastate conditions, and the *Hermod* could be made available at the time required.
(3) *Installation aids.* On the UMC there were five rigging/ballasting platforms to be above normal wave height. Four lowering padeyes and pre-installed 112 m diameter × 86 m long nylon slings, pre-installed warping/towing lines at each corner and draft markings to assist in ballast calculations.
(4) *Instrumentation.* Also pre-installed on the UMC were instruments for precise determination of level and bearing. To allow precise location these were referenced to a transponder array placed under the control of in-field surface-positioning references.
(5) *Communication/training and safety.* Efficient communications, personnel training and safety were stressed throughout the programme.
(6) *Contingencies.* A full complement of contingency measures was defined. These covered the possibilities of problems during tow,

*1 ton ≈ 1.016 tonne.

during launch or surface-handling, and after placement on the sea bed.

(7) *Pre-launch check-out.* A very thorough check-out of all of the UMC equipment and fixtures was made before tow-out and launch.

Load-out for the launch was performed, by a specialist contractor, using platform trailers on a total of 800 wheels with pneumatic tyres. Actual movement on to the barge was done comfortably within one 4 h period of high tide at Schiedam, The Netherlands.

The 560 nautical mile tow began on schedule to rendezvous with the crane vessel some 3 days later at Central Cormorant. Weather conditions of Seastate 5 or less were met, and the tow was uneventful. Upon arrival at the Central Cormorant location the UMC was thoroughly checked-out against a comprehensive checklist and, being undamaged, was found suitable for launch. The seastate was lower than the limiting 1.5 m significant wave height and, with a favourable forecast, preparations were begun for launch. The sea-fastenings were cut and the barge was ballasted to attain an attitude of about 4° down by the stern. A haul-off force of approximately 200 tonnes was then applied to induce launch. After launch the UMC was towed to the stern of the *Hermod* and manoeuvred under the crane by two tugs. Pre-installed nylon slings were connected to the 500 ton block. Two 40 mm diameter nylon warping lines were also connected to the 500 ton block, and two 40 mm nylon warping lines were attached to tugger winches on the *Hermod.*

Positioning lines from under the roll of the *Hermod* were attached to the UMC. Positioning equipment, pre-installed on the UMC, was connected by electric/communication umbilicals to the *Hermod.* All systems were checked for satisfactory operation, and the UMC was ready for ballasting and lowering.

Draft marker readings on the UMC were read to confirm the calculated weight and buoyancy. Ballast and trim tanks were then selected for flooding from the 44 available tanks. An in-house developed computer program, for UMC weight and buoyancy control, was used to give the UMC an approximate 85 tonnes negative buoyancy. The appropriate valves were opened, by riggers, to flood the selected tanks completely.

The ballast tanks filled quite slowly and in approximately 1.25 h the UMC submerged. As the unit submerged it gradually extended the nylon lowering lines which had been relieved of slack but were not taut. The UMC was lowered at increments of some 2–3 m/min until near the sea bed. During the lowering the hook load increased up to (measured) 80 tonnes as the ballast tanks were completely filled. After checking adjustments to positioning equipment and confirmation of bearing and position, the UMC was lowered gently to the sea bed. Position and bearing level were checked and found to be acceptable.

North-East Frigg/SKULD System

North-East Frigg control. Field control station (FCS). An articulated column installed 150 m from the subsea station supports the field control station (FCS). The FCS houses the equipment required for:

- converting electrical control signals from the Frigg Field into hydraulic pressures for operating the subsea gas production valves;
- closing control of wells through individual 2 in. kill lines;
- periodic leak testing of the production tubing safety valves in each well;
- continuous injection of hydrate inhibitor into wells during gas flow.

Umbilicals. Six specially designed umbilicals, each consisting of one 2 in. tube and twenty ½ in. tubes, link each Christmas tree to the deck of the FCS.

Links to the Frigg Field. The North-East Frigg facilities are linked to the TCP2 Frigg Field platform via a 16 in. gas line. A 1 ¼ in. hydrate inhibitor supply line and an electric cable provide the FCS with both electric power and remote-control signals. These signals are also transmitted through a back-up radio link. Treatment, metering and boosting modules are installed on TCP2. This allows the North-East Frigg gas, after treatment, to be injected into the Frigg gas flow.

SKULD system. Elf-Aquitaine Norge A/S planned to develop subsea remotely controlled clusters, as part of the SKULD programme. The programme was detailed in two phases:

- Phase 1. Qualification of a subsea production remote-control system;
- Phase 2. Manufacturing of complete subsea systems for marginal fields in the North Sea.

Main objectives of the project

(1) Produce over a long distance (20 km) between subsea cluster and control room.
(2) Diverless installation and retrievable control system.

(3) High reliability (5 years to a shutdown).

Main principles

(1) Use of Grondin's experimental results.
(2) Trials with two Christmas trees: a live one (the production tree in this case is simulated) and a dummy.
(3) Use of either qualified or field-proven products.
(4) Electro-hydraulic control system.
(5) Electronic multiplex transmission.
(6) Closed hydraulic circuit using subsea pumps, with hydraulic power supply from the platform in back-up.
(7) Back-up for electric power supply and transmission through a buoy.

Control system. The SKULD experimental subsea station is located 1 km offshore in 100 m of water. The control room is onshore, with back-up via the buoy. A general view of the experimental station is shown in Fig. 12.

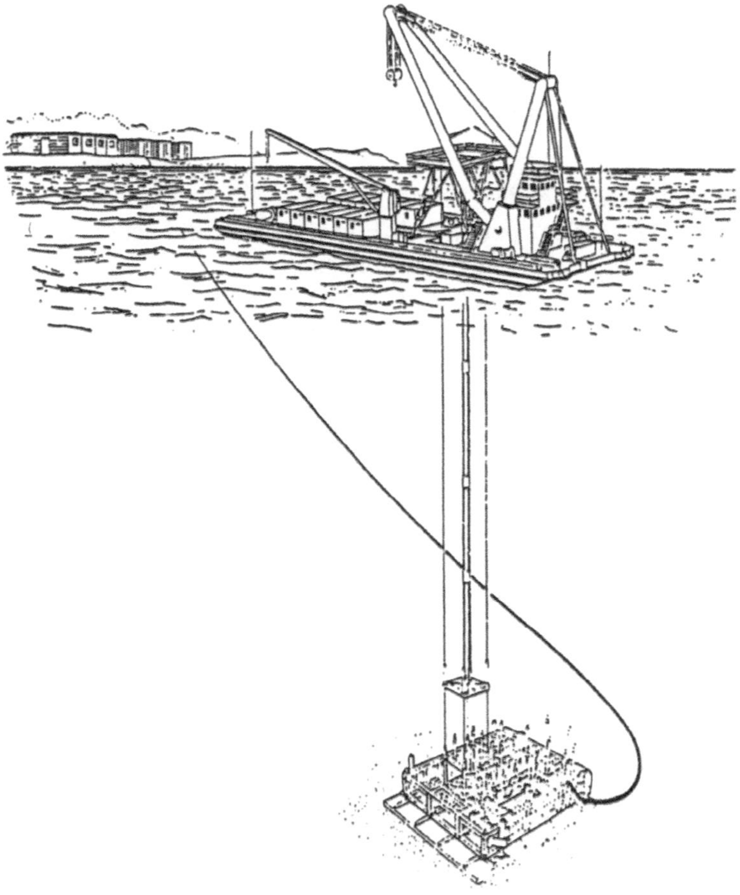

Fig. 12 Elfnorge SKULD experimental station.

Subsea station. The subsea station consists of two wellheads on a template and a manifold controlled through an electro-hydraulic system. A general view of the subsea station is shown in Fig. 13. The items on the template manifold are:

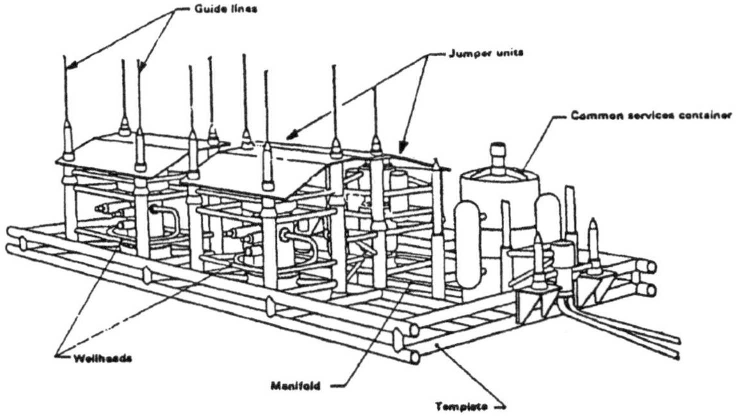

Fig. 13 Elfnorge SKULD subsea station.

- two subsea wellheads;
- a common service container (CSC) with a bundle head;
- two production and control jumper units;
- a connecting frame (to connect CSC and jumper units).

On each tree, the following functions are remotely controlled (eight commands):

- six fail-safe valves;
- two DHSVs.

Also, pressure measurements, valve position indicator and gas detectors. On each jumper unit (three commands):

- two fail-safe valves;
- one choke.

Also, redundant pressure measurements on each line (flowline, annulus line and methanol injection line), temperature transducers, valve position indicators, gas leak detectors, hydraulic fluid pressure and flow measurement, solenoid valves position indicators.

On the manifold (nine commands):

- six fail-safe valves;
- two close-assist valves;
- one bowl valve.

The common service container (CSC) consists of subsea hydraulic power supply, control and monitoring. Control and monitoring cover subsea pumps start-up, reservoir management, hydraulic pressure measurements, flowmeters, water and gas detection, and solenoid valves redundancy management.

Electric supply control and monitoring covers installation measurement, platform/buoy switching, solenoid valves position indicators and subsea/onshore hydraulic power switch monitoring.

To connect the several packages together, automatic connectors are provided through what are defined as "multiconnectors".

Surface equipment. The SKULD electronic surface unit consists of:

- control desk equipped with two-colour remote-control terminals and a functional keyboard;
- central processing unit with a minicomputer system and associated interfaces.

The subsea remote terminal units (TESUMA) are connected to the central connecting unit via two alternative transmission links:

(1) hard-line transmission through the main platform umbilical;
(2) radio transmission from the main field to the buoy, then a hard-line transmission from the buoy to the cluster.

All of the subsea information (valve positions, hydraulic status, pressures, gas detection, etc.) are locally gathered by the remote terminal unit. The information is then transmitted upon request via one of the transmission links to the central processing unit in the form of serial data bits (scanning cycle). This information is processed and presented to the operator at the control desk. The operator is automatically alerted of all alarm conditions at the subsea level on critical events. Shutdown actions are automatically elaborated by the system.

From the control desk the operator can operate the valves and pumps using the remote-control terminal (CRT) and the functional keyboard. Corresponding local actions are completed by the remote terminal unit concerned.

Also, all surface information (electrical power status, hydraulic or methanol production status, etc.) is gathered by the system and displayed on the control desk.

Maintenance in "Grondin" Field (TIM robot)

The Elf-Aquitaine group has conducted an extensive research and development programme to develop subsea production for extreme water depths beyond the divers' capability (600 m).

The programme has included floating drilling and completion of highly deviated wells. Also in the programme are the setting and connection of subsea structures without diver assistance, remote production of subsea wells and automatic flowline connections. Remotely operated vehicles (ROVs) were used for intervention and maintenance, and for reliability and safety analysis.

An experimental station has been located on the north-east extension of the "Grondin" Field, offshore the Republic of Gabon, West Africa. Located 1500 m from the fixed production platform complex, the station is installed in 60 m of water.

The barge *Anguille* was used as a support vessel with surface equipment. Subsea equipment included template, Christmas trees, production manifold and the flowlines. All of the equipment is remotely controlled from the *Grondin* production platform.

Remote control was via a subsea electrical cable with local electro-hydraulic subassembly and surface monitoring facilities. Installation, connection, maintenance and diverless operations were conducted from a floating service vessel designed and constructed specifically for the station.

From initial engineering in 1974, the *Grondin* north-east station marked a first step towards establishing the feasibility and reliability of subsea production in deeper waters.

In 1981, Elf-Aquitaine completed its successful remote controlled subsea production system tests. The final phase of the test involved the use of a rail-mounted remote control robot manipulator developed by ACB (Fig. 14). Following 6 weeks of testing of the 13 tonne remote-controlled manipulator TIM (Télé Manipulateur d'Intervention et de Maintenance), two modifications to the system were proposed to increase the capability of handling production from 600 m water depths. These proposed modifications first involved substituting an electronic multiplex control system, for the electro-hydraulic system to reduce weight. Secondly, the upgrading of the visual system (mimic diagrams) used to show the work being performed on the subsea production system to the topside operators (see Fig. 15).

Controlled by operators at a console located on the support barge *Anguille*, TIM successfully carried out 104 different operations. These operations were carried out during 24 separate dives which involved a total of 160 h on the bottom. During tests the system's

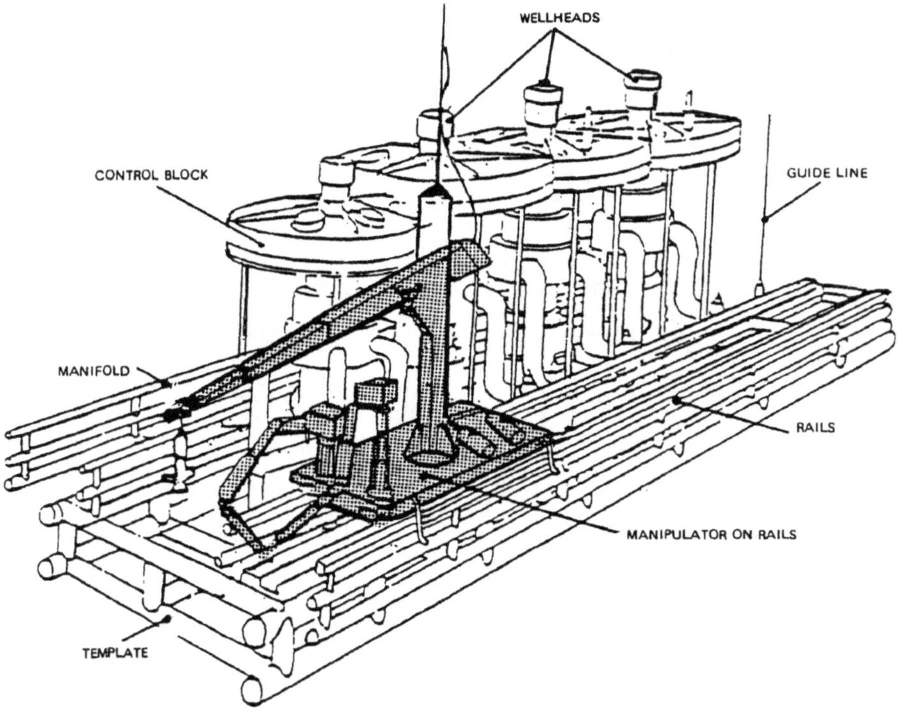

Fig. 14 Elf Grondin template trees and rail-mounted robot.

manipulator was brought on to the barge between trials for tool change-out. However, this work could be done underwater, with the robot selecting tools needed for its tool store and performing long periods of maintenance on the sea floor.

Although the depth at the experimental site was only 60 m, the design limit of 600 m was selected for the experimental performance of the manipulator package. This criterion imposed care in selection of the components to be immersed as well as a conceptual study excluding any direct human intervention.

Missions of TIM robot. (1) Connection of hydraulic jumper pipes between a wellhead and a flowline on the manifold.

(2) Hydraulic supply for operation of connectors.
(3) Laying of electric cables and hydraulic hoses.
(4) Assistance for electric jumper cable positioning and connection.
(5) Installation of operating tool for valve opening or closing (manual safety valves on manifold and wellheads).
(6) Releasing and connecting the guide lines.
(7) Brushing, grinding, cutting, tightening and loosening nuts and bolts.

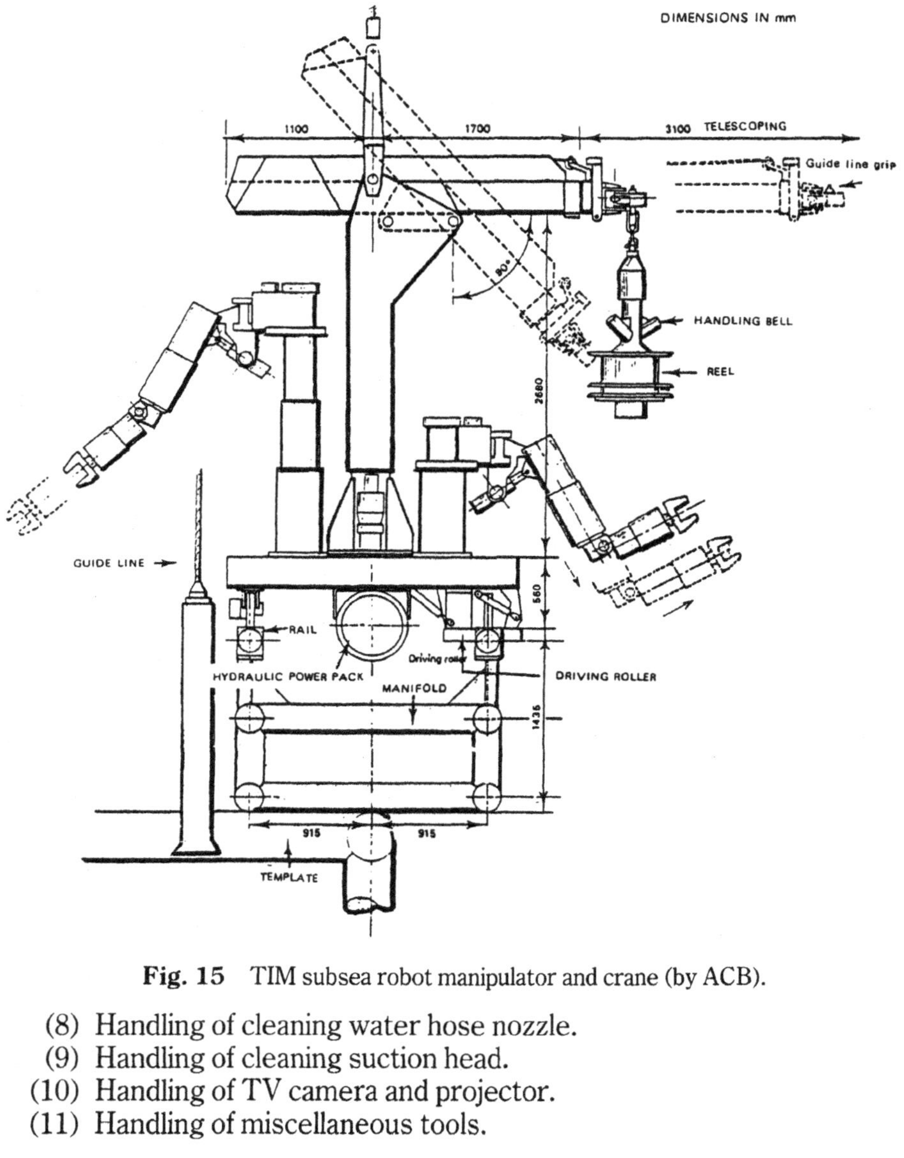

Fig. 15 TIM subsea robot manipulator and crane (by ACB).

(8) Handling of cleaning water hose nozzle.
(9) Handling of cleaning suction head.
(10) Handling of TV camera and projector.
(11) Handling of miscellaneous tools.

Maintenance Shell/Esso UMC

General. The UMC is designed for three primary modes of maintenance:

(1) Drill ship – for well-related work, piping spools and some pipeline connectors.

(2) Remote maintenance systems (RMS) – for maintenance by replacement of manifold valves and control modules.
(3) Diver interventions – for back-up for RMS, assistance with drill-ship maintenance and to perform miscellaneous "one-off" tasks.

Inspection is planned with remotely operated vehicles (ROVs), divers and, as a secondary support, the RMS vehicle (see Fig. 16). A system of markings has been used on the UMC for identification. Provisions for maintenance have been made based upon design criteria and equipment test results. Actual maintenance requirements have also been determined through operating experience and the results of inspections.

Drilling rig operations and template guidance system. Rig run tools may be used to retrieve UMC equipment items for surface maintenance as required. During such operations the rig is located above the UMC installation and performs operations using special tools provided. These tools provide the capability to perform the following operations:

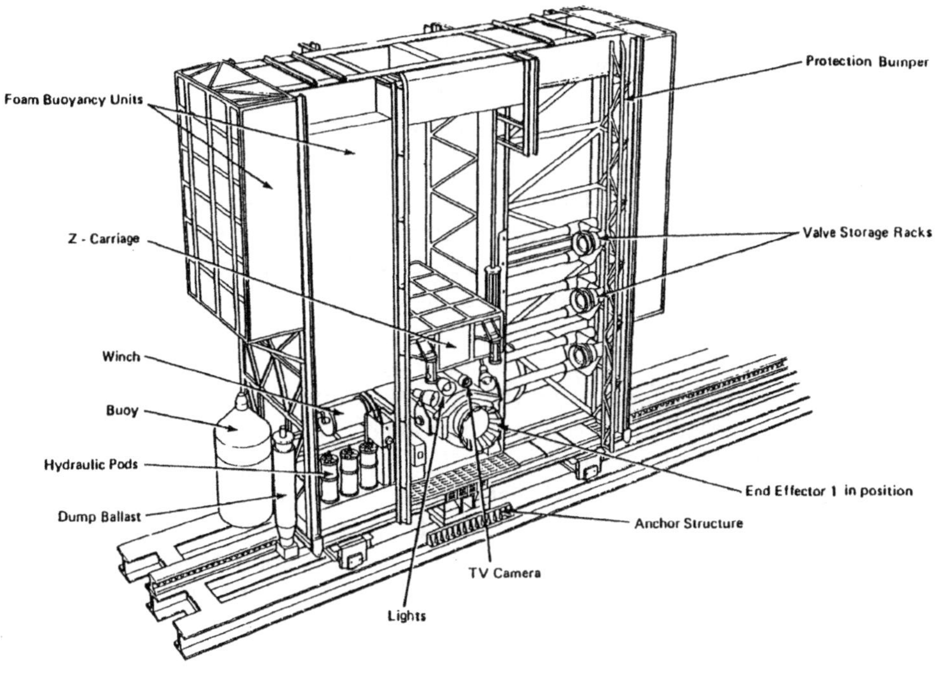

Fig. 16 UMC maintenance – RMS details.

- remove or replace satellite well choke spools on the UMC;
- remove wellbay roof and open Christmas tree for workover operation;
- remove hydraulic accumulator unit or chemical injection package units for repair or replacement;
- pull-in and connect flowlines, electric and hydraulic control lines;
- install aft RMS landing area or remove and replace areas of RMS track;
- install underwater shutdown connector or rig electrical connector as back-up facilities or for system diagnostic checks.

In order to perform these operations the rig utilizes the template guidance system to locate equipment accurately on the template. There are 42 locations in the UMC requiring guidance and some 122 guidepost receptacles. Guideposts may be established in the receptacles using purpose-made tools. A re-entry guideline may also be established to the guidepost.

Remote maintenance system (RMS). The RMS includes a robot manipulator vehicle. The RMS is designed to perform the following specific tasks:

- remove and replace insert valves;
- operate isolation valves;
- remove and replace control modules;
- undertake remote camera inspection.

The features of the vehicle are shown in Fig. 16. The vehicle is built within a main structural assembly, foam filled for buoyancy, and tall enough that tools may be operated to any height required on the manifold. The frame includes a storage rack for replacement parts. Operations are performed using two main tools – end-effectors. End-effector one (EE1, a socket wrench) is used for valve operations. End-effector two (EE2, dual torque wrench and fork lift) is used for replacement of control modules. The tools are operated and positioned hydraulically.

Television cameras and lights are mounted on the vehicle, therefore operations can be viewed from the surface control panel located on the support vessel during maintenance operations.

There are two RMS landing areas, positioned fore and aft of the UMC installation. Both landing areas are detachable, with only the forward area in place during the launch and the aft landing area installed later if required. On deployment the RMS is landed on to the UMC and moves along the track to perform operations on equipment located along the main manifold pipework.

TEXACO HIGHLANDER SLUG CATCHER

Design philosophy

The Highlander Field is located in Texaco's Block 14/20 in the UK sector of the North Sea, some 8 miles West of the existing Texaco Tartan Field and approximately 100 miles north-east of Aberdeen, as shown in Fig. 17. In 420 ft of water, the field is considered marginal within North Sea development parameters, with an estimated 35 million barrels of recoverable oil.

In December 1974 Texaco discovered the "Tartan" Field in Block 15/16 of the UK sector of the North Sea. Subsequently, Texaco installed the Tartan "A" platform at that location. Tartan "A" is a fixed steel structure fitted with a complete topside facility for processing and transporting crude oil and gas to shore via pipelines. However, oil production from the Tartan Field has not reached the original 75 000 barrels/day of oil throughput design of the facility. Consequently, excess capacity existed to handle, with certain modifications, production from other nearby fields which awaited development.

A potential recoverable reserve of 35 million barrels was calculated for the accumulation, which is considered marginal in size by North Sea standards. The quality of the oil is high, but the gas-to-oil ratio is low, with correspondingly low reservoir energy. Therefore, water injection and artificial lift are both ultimately expected to be required to maximize recovery.

Evaluation of the discovery well led to the drilling of three further delineation wells and one water injector. These wells are directional-

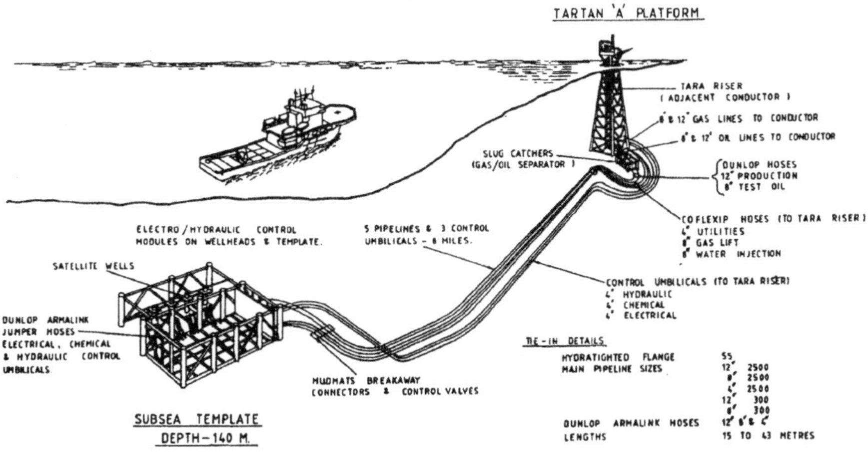

Fig. 17 Texaco Highlander Field. Flowlines, umbilicals and flexible spools.

ly drilled from wellheads placed in a line approximately 25 ft apart. With encouraging results from the discovery well, this configuration was chosen at an early stage as one which could later be adapted for satellite well production through an adjacent template. All of these wells were extensively tested and suspended for future production or water injection. During this evaluation phase, pulse tests were conducted providing positive communication existed between three of the wells.

Four systems were considered for development of the field, each of which disposed of oil and gas through pipeline systems terminating at the Tartan platform. These systems, with their relative project investment costs, including wells, were as follows.

	Relative cost
Fixed self-contained platform	2.9
Fixed satellite platform	2.8
Purpose-built semi-submersible floating facility with template	2.5
Remote subsea production facility	1.0

The substantially lower investment of the subsea facility and the fact that production could be achieved much earlier from the four existing wells both coupled to make this marginal field project economically viable. Texaco's first-hand experience with satellite subsea wells and their associated control systems was deemed sufficient to ensure the technical success of a subsea facility.

The design philosophy and construction approach for the facility was governed by the following basic criteria:

(a) provide a simple low-cost but operable and reliable facility to maximize profitability;
(b) utilize proven equipment and hardware wherever possible;
(c) avoid experimental systems;
(d) anticipate diver access for installation and maintenance;
(e) conduct thorough pre-commissioning and checkout before installation;
(f) minimize interruption to the Tartan operating time during installation;
(g) minimize weight addition to the Tartan platform;
(h) attain early production from the existing wells at the beginning of 1985 and complete the subsea facility installations during the summer of 1985, in less than 2 years from the start of engineering;

(i) take advantage of current saturation-diving capabilities for installation and future maintenance.

The Highlander Field is characterized as a relatively low energy reservoir with an inherently low gas-to-oil ratio (GOR). Initial testing has proved the potential for an improved recovery utilizing water injection. Further artificial assistance was required to supplement the natural reservoir energy, particularly under conditions of increasing water cut. Texaco decided to adopt conventional downhole techniques for gas lifting, but were consequently faced with having to solve the complexities of distribution and control of lift-gas demand at a template 8 miles distant. The resulting high GOR and consequent pipeline slug flow presented additional challenges in delivering the crude and gas to the Tartan "A" Topsides without production-inhibiting back-pressure on wells or topsides space/weight penalties.

Analysis of crude from the development wells indicated good-quality light oil, but with wax and CO^2 contents that could not be ignored while considering conceptual designs and material specifications. A pour point and hydrate formation temperature higher than sea-water temperatures also required careful consideration of numerous potential methods of control.

To combat the high pour point of the crude, solutions involving insulated and/or looped pipelines were investigated and rejected in favour of continuous injection of pour-point depressant (PPD) chemicals backed-up by a subsea pigging system. Internal corrosion problems were dealt with by a combination of carefully chosen metallurgical solutions (in manifolds and valving) and injection of film-forming chemical inhibitors (for pipelines, slug catchers and other downstream devices).

The Tartan "A" topsides revamp design work was carried out under the strict restraints of minimizing interruptions to Tartan Field production and weight addition to the platform.

The duty and capacity incorporated into the system design is summarized in the following section.

Subsea slug catchers and pumping system

Once the crude arrives at the base of the Tartan platform 8 miles from Highlander, slug flow will have developed fully in bulk production and test pipelines. The design challenge at this point was to provide full throughput from the low-energy Highlander Field to the topsides some 550 ft above the sea floor without causing process upsets due to severe slug flow, and without increasing weight on the

Tartan jacket. The solution developed from Highlander was to place slug catchers on the sea floor adjacent to the base of the jacket, as illustrated in Figs 18 and 19. The slug catcher performs the dual function of smoothing the slug flow from the pipelines and allowing

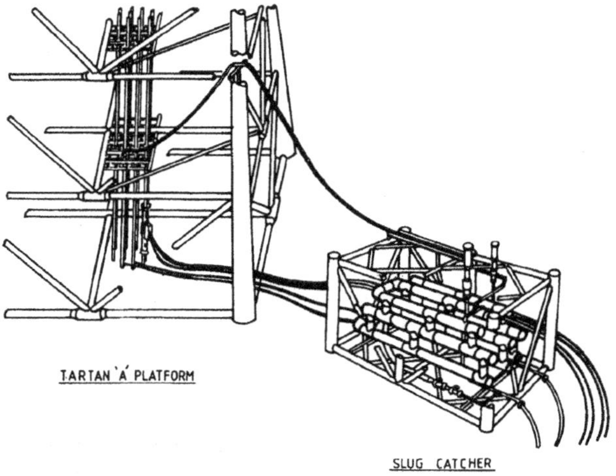

Fig. 18 Slug catcher at base of Tartan platform.

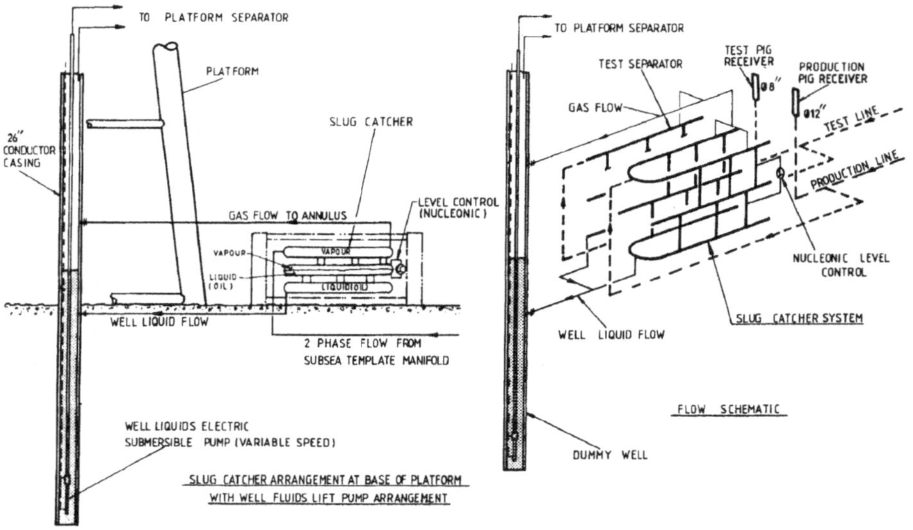

Fig. 19 Slug catcher and dummy well arrangement.

gas and oil to be separately delivered to the platform topsides without depending on wellhead pressure for the motive force. This was accomplished by the design of a three-tiered configuration slug catcher for each system and the use of electrically driven downhole pumps contained in two self-supporting risers installed in spare drill slots in the Tartan "A" platform. The bulk production slug catcher is three-tiers high, two bays wide, and the test slug catcher is three-tiers high and one bay wide.

The slug-catcher chambers are constructed of 36 and 48 in. diameter grade 52 pipe and fittings, and are built to pipeline codes without any external penetrations. The slug flow in the pipeline is delivered to the top tier of the slug catcher, which serves as a chamber for separation of the oil and gas. The gas is taken off the top of this chamber and delivered through catenary flexible pipes to the annulus of the riser. The catenary configuration provides flexibility of differential movement between the two structures while preventing any pocketing of carry-over liquids in the gas. The oil drains to the centre chamber of the slug catcher, which alternately fills and drains, acting as a surge chamber to catch the slugs and smooths the total throughput to the risers. The bottom chamber of the slug catcher provides a horizontal storage volume of liquid, preventing the slug catcher from running dry, thereby allowing the level in the vertical riser to remain essentially constant. Fluid level detection in the slug catchers is accomplished through the use of an array of nucleonic density switches mounted on closed-loop bridles on each slug catcher. Signals from these switches are transmitted via cable to a microprocessor-based controller which activates the flow control valve on the Topsides discharge piping of the pump. This microprocessor has a resident database which is continuously updated with slug-flow information. This is used in conjunction with a specially developed algorithm to maintain a constant flow throughput to the topsides separators while the slug catcher fills and drains.

The crude flows out of the slug catchers via flexible pipes to side inlet flanges on the 18 ⅝ in. riser casings, then down the annulus to the submersible pumps located approximately 1000 ft below the sea bed. The pumps are suspended on 7 in. tubing strings which carry the pumped crude to the topsides. The gas from the top of the slug catchers enters the riser annulus via side inlet flanges located above the liquid level, and flows upward to the surface process equipment. In addition to the flow control valve at the topsides, a minimum-flow bypass valve allows the pumps to run continuously, even in low-flow conditions. High pump reliability is anticipated because the downhole pumps are designed for severe duty in deep, deviated holes pumping hot two-phase crude against a variable head, whereas Highlander operating conditions provide a shallow vertical hole pumping cool single-phase crude against a constant head.

Through the use of variable-speed drivers and the minimum-flow bypass valves, these pumps are expected to remain in continuous duty for long periods, therefore avoiding premature failure due to frequent stopping and re-starting. When a re-start is unavoidable, a "soft" start is allowed using the variable-speed equipment. The overall dimensions of the structure are shown on Fig. 20.

OPERATING COSTS

Introduction

The option chosen for evaluation of operating cost is the floating platform system with offshore loading, which is selected for marginal field development in the North Sea. Using a semi-submersible, as a production facility and subsea completions, defines the smallest economical field to be developed independently.

This section sets out the breakdown of operating costs in 12 cost centres, based on expertise and various financial studies already carried out. For each subject treated, the cost of personnel involved have been taken into account. The costs are summarized in Table III and described in the following sections.

Increased competition

Underwriters have also felt increasing pressures from new capacity brought into the market by up-and-coming competitors in Japan, Europe and Bermuda. Therefore, according to one industry broker, there is more than ample capacity today. This situation will probably continue for the next 2 or 3 years.

Major insurance losses

As rates have been dropping, however, marine accidents have been increasing. The growth in offshore drilling, coupled with harsher operating conditions, has brought about a surge in mobile-rig accidents in recent years (*Offshore*, March 1981). An earlier report on offshore accidents showed a monetary loss involving 140 mobile-rig accidents totaling US$744 million over a 26–year period, not including nine accidents for which damages have not been determined.

Almost 50% of total losses over this period occurred in the period 1979–80, which involved the damage of 22 rigs and total loss of eight others in a 2-year period. Total losses were US$356 million in damages, not including six mobile rigs for which damages were not assessed.

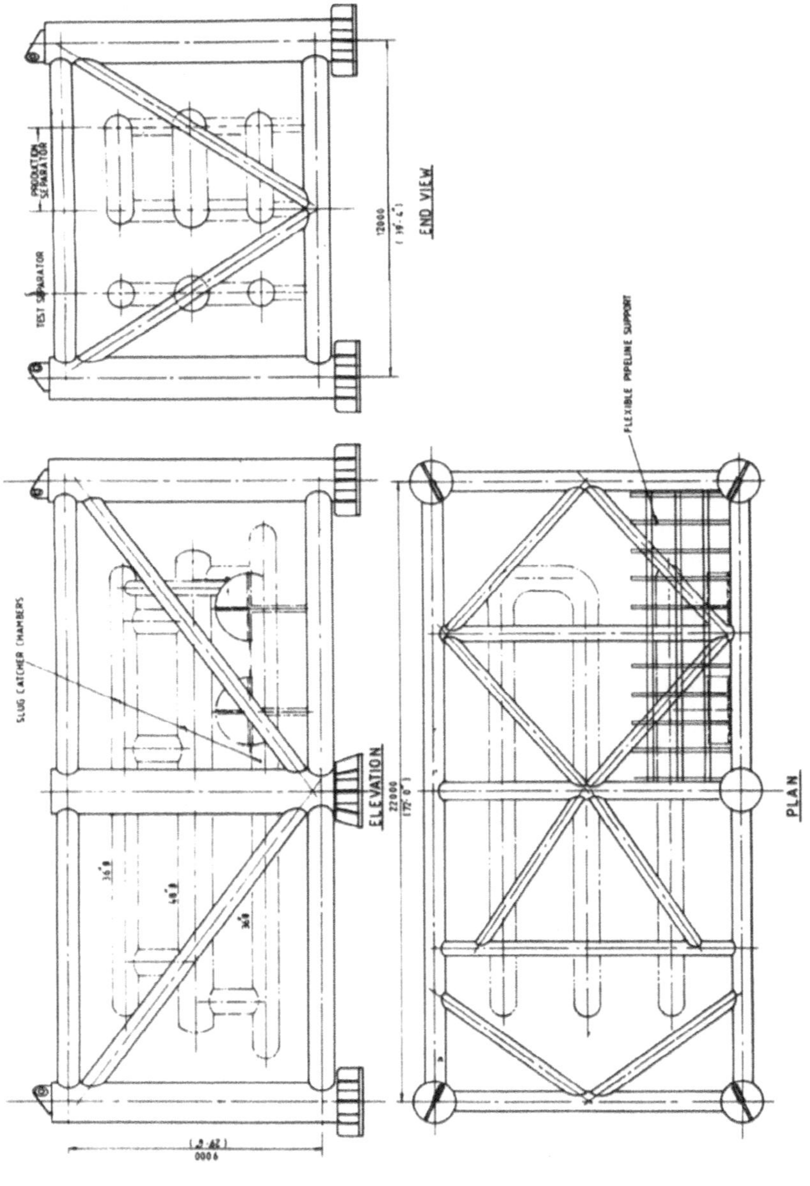

Fig. 20 Slug catcher, general arrangement.

Table III
Breakdown of operating costs

	Cost centres	*Percetage of operating costs*
1.	Insurance	20
2.	Transportation	12
3.	Underwater IMR	22
4.	Topside IMR	
	General work	9
	Processing	3
5.	Logistic	5
6.	Consumables	4
7.	Administration/management	10
8.	Safety (training)	1.7
9.	Communications	1.3
10.	Miscellaneous	12
	Total	100

Transportation costs

Based on the cost of shuttle tankers, 12% of operating costs for transportation (*International Petroleum Times*) is considered a maximum figure. This can be possibly brought down to 9–10% when commercial circumstances on the tankers market allow.

Underwater IMR

Inspection costs

Research shows that chief among the items of expenditure on inspection, maintenance and repair (IMR) is that of underwater inspection and maintenance, with 22%. This is the biggest slice of the business. Part of the reason for the relatively high cost of inspection and maintenance underwater is the difficulties it entails. Offshore structures, especially the earlier installations, were not designed for ease of inspection access, either by ROV or by diver, which can make investigation time-consuming and inefficient, not to mention hazardous.

Research and development

Much research and development time has gone, and will continue to go, into evolving quicker and more-efficient ways of monitoring the integrity of structures and of improving their inspection. Similarly, operators are now increasingly concerned with designing in structural maintainability, at both the routine and the more major levels. However, it will take time before its effect can really be seen in the market, and the money spent on such research will also have to be taken into account. Consequently, it is felt that the IMR market will continue to increase in value.

Structural maintenance

Structural maintenance and combating corrosion, cleaning, painting and application of coatings, take up an important part of this. This means each of these elements could, by 1987, singly provide an annual market worth more than £150 million. Even during this year, 1984, the total value of each of these areas will be more than £100 million.

Total expenditure

Given the immense sums expended on field development, the value of the inspection, maintenance and repair market is hardly surprising. In the North Sea, for example, during 1983 capital investments in field development were £1985 million in the UK and Nkr 24 000 million in the Norwegian sector. As a general rule, operators can expect to commit to expenditure of about £10 million per year on operating facilities (production and transportation) out of every £100 million of capital expenditure.

Of each £10 million of running costs, approximately 22% will be for inspection, maintenance and repair. As technical complexity increases to meet the field development challenges of the future, development costs must also increase, and with them the value of the offshore IMR market.

Topside IMR

The cost of general work in IMR of the topside is estimated at 9% of total costs.

The cost of process operations is estimated at 3% of total costs. These figures cover the cost of topside equipment and maintenance, current engineering work, piping, electrical and mechanical inspection.

In the figure for general work, the cost of spares is included, which is generally estimated at 3–5% of annual operating costs.

Logistics

The cost of logistics includes the following items:

- helicopters;
- workboats;
- yard and warehousing facility;
- harbour facility.

All of the above rates are clearly dependent on the market. However, a reasonably precise figure can be assessed. This takes into account:

- supply ship rates;
- helicopter rates;
- yard and warehousing cost per square foot;
- harbour rates.

The cost is estimated at 5% of the total.

Consumables

Consumable costs are inclusive of:

- offshore catering (food, laundry and cleaning);
- lubricants and chemicals;
- waste disposal (return to shore).

The estimated cost of consumables is 4% of the total.

Administration/management

Administration and management consist of:

- management and certification;
- onshore administration cost.

The estimated cost of these is 10% of the total operating expenditure. With improved cost control it should be possible to reduce these costs.

Safety

Safety activities consist of the following:

- fire-fighting;
- safety training;
- evacuation lifeboats;
- oil-spill control.

The percentage costs of the total is 1.7%. This relatively low figure shows that it is operating costs and maintenance of equipment which are of fundamental significance in the cost structure.

Communication

Communication costs include the following:

- offshore/onshore links;
- satellite communications;
- telephone.

The cost of communications is 1.3% of the total.

Miscellaneous

For forward planning and budgeting this cost centre covers the provision of contingencies. After expenditure the costs can be either re-allocated or listed as miscellaneous.

With the uncertainties of planning and organization of offshore and underwater operations, the miscellaneous cost centre (contingencies) are high (12%) compared with onshore operations.

Unscheduled occurrences for which costs can be incurred (e.g. repairs) relate to both topside and underwater. Underwater repairs, due to location and access, are more costly and North Sea underwater repairs on structure and pipeline have ranged between £2.5 and 9 million.

Table III gives a summary of the cost percentages for each item described above.

SUBSEA WIRELINING

Early production

Wireline operations are normally carried out from the surface vessel, drilling rig or floating production vessel, through the workover/completion riser (refer to Fig. 21 for more details). The wireline unit is mounted on the surface vessel, inside the derrick. Wireline

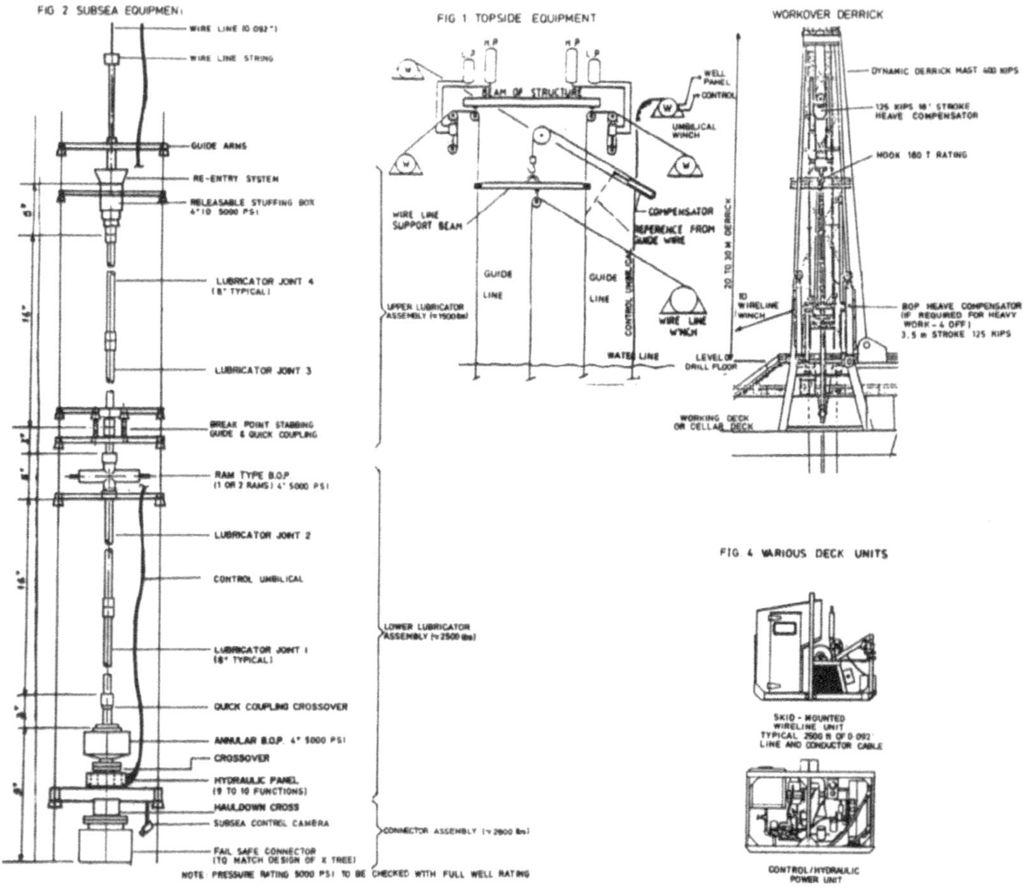

Fig. 21 Subsea wireline operations.

operations have been carried out by Flopetrol-Schlumberger, first in Zakum Field, and then in various oilfields around the world. These used subsea wireline lubricator and associated equipment, keeping the wireline winch, hydraulic power unit and remote-control panel on the support vessel. The operations have been very successful. However, Flopetrol-Schlumberger have not achieved this in the North Sea.

North Sea

Improved technology permitted another company, Otis, in conjunction with Comex Houlder, to carry out a successful test for Hamilton Brothers in 1984. This was in the North Sea. The objectives of the test on a real wellhead were:

- position-keeping ability of a medium-sized DSV-type vessel;
- lubricator make-up, rigging and handling methods on the deck without derrick, but only a heave-compensated crane;
- subsea handling and installation with ROV and/or divers;
- wireline system release and retrieval;
- ultimate development of a method of performing wireline work from a dynamically positioned diving support vessel (DSV with DP).

The method used avoided the use of guide wires, but this is not recommended. The DSV used was Seabex one, a medium-sized modern DSV of 6000 tonnes displacement. The test was successful, but the need for a handling derrick was strongly demonstrated.

Typical deployment operation

A typical deployment operation from a DSV is shown through five key figures in Fig. 22, and typical equipment required on deck and subsea are presented in Fig. 21.

Operations carried out by subsea wireline can be:

(a) running and pulling the wireline-retrievable DHSV (depending on completion design);
(b) bottom-hole pressure survey;
(c) bottom-hole temperature survey;
(d) production logging;
(e) tubing services, workover only;
(f) replacement/servicing of gas lift valves.

ROV/diver assistance

The only sequences of operation requiring ROV/diver assistance are:

(a) setting of guide lines;
(b) removing Christmas tree cap protector;
(c) installation of special flange for hydraulic connection.

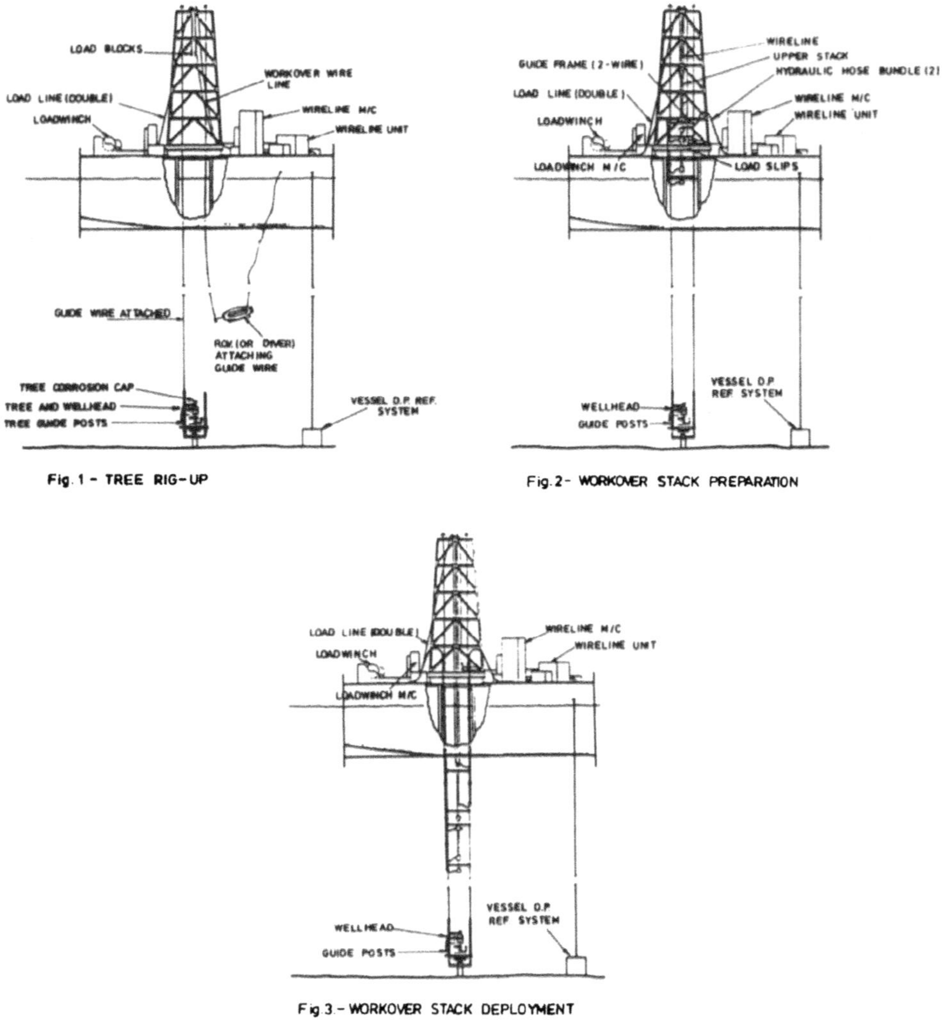

Fig. 22 Typical deployment from a support vessel (DSV).

Releasable stuffing box

The special releasable stuffing box allows various sequences to be performed without having to pull the complete lubricator and BOP assembly. The subsea lubricator can operate either with slick line, for wireline operation, or with a conductor cable, for logging of pressure read-out. Typical dimensions and weight of equipment are indicated in Fig. 21.

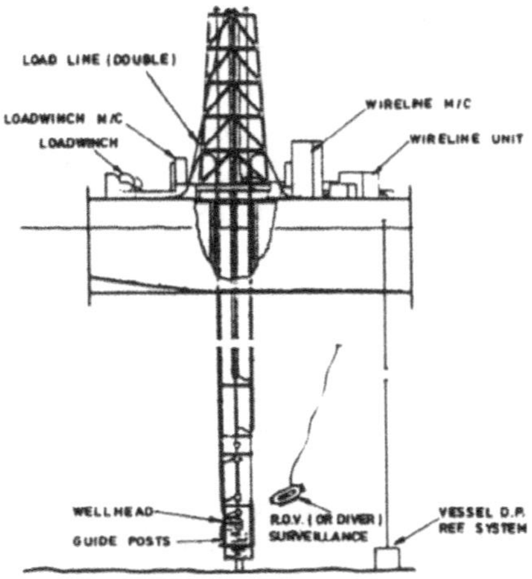

Fig. 4 - WORKOVER STACK DEPLOYMENT-LANDED-RETRIEVAL TOOL DEPLOYED

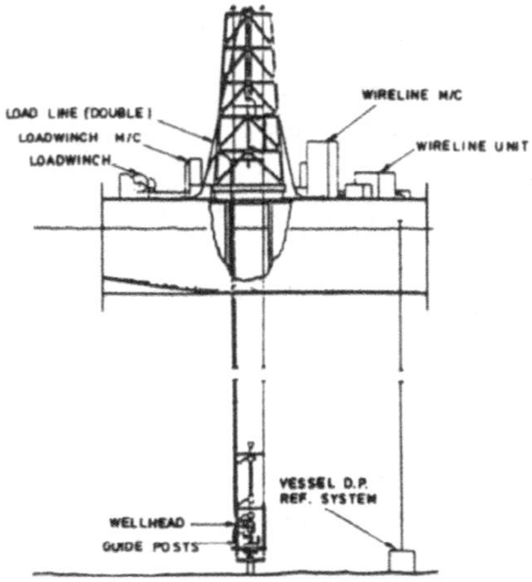

Fig. 5 - WELL WORKOVER-TOOL REPLACEMENT IN UPPER STACK

Fig. 22 (cont'd)

Subsea wireline BOP

It is becoming increasingly common for North Sea operators to turn to the use of coiled tubing for workover operations to solve many common well problems. Many of these problems may be solved using alternative methods; however, compromises usually have to be made to achieve satisfactory results.

Most operators, when running coiled tubing from a floater, have a subsea coiled tubing/wireline BOP in addition to a surface BOP. This is felt to be a necessary precaution in the event of an emergency occurring with the coiled tubing in the hole. The subsea tree valves could not be closed to secure the well with the coiled tubing across them. With a subsea BOP the shear rams can cut the coiled tubing, allowing it to fall clear of the tree valves, which permits the upper section of coiled tubing to be pulled clear.

Some of the problems that may be tackled using coiled tubing methods are:

- kicking-off difficult wells;
- back-flushing water injectors to clean them up;
- spotting acid or scale treating chemicals across a specific zone;
- circulating scale/fill from top of a stuck wireline plug;
- circulating a tubing retrievable subsurface safety valve should valve fail – this may dislodge the debris causing the valve to fail;

- spotting cement or water shut-off chemicals should one particular section of a production zone be producing all of the water;
- sand clean-outs, should any of the gravel packs fail;
- other treatments or stimulations that may be required, should satellite wells be brought on-stream from different reservoirs.

Unfortunately, previous use from floaters has been somewhat restricted due to a lack of an efficient means of heave compensation. Recent experiences with the system have shown that this problem can be overcome. In addition, further work is being carried out using integral risers fitted with a coiled-tubing injection head. In this instance the heave-compensation requirements are catered for by the workover riser arrangements through tensioners and the heave-compensated derrick to which the handling/terminal head is attached. The tubing head replaces the conventional wireline BOP and lubricator head.

Areas for improvement

Wireline operations

Wireline operations, using the conventional lubricator mounted directly on the wellhead plus a swell-compensating device, have proved feasible and perfectly practical in limited water depth. In deeper water it will be necessary, however, to develop alternative means of carrying out wireline work. The simplest and most adaptable method evaluated in the Zakum scheme in the 1970s appeared to be the remotely controlled submersible wireline winch. However, more development work would be required to bring this to a reliable and useful state.

Deeper water

For use in deeper water (where divers or guidelines are not considered) it would be necessary to operate the winch with a search propulsion system in order to locate and move to the wellhead. A remotely operated connector for latching on to the wellhead can be made available, and adequate internal power and electronic capability would accommodate the propulsion and search systems.

North Sea

A subsea wireline is apparently being developed in Aberdeen for North Sea operations. Hamilton Brothers, again, have expressed interest and a test was planned for 1983 (refer to Fig. 23 for more details).

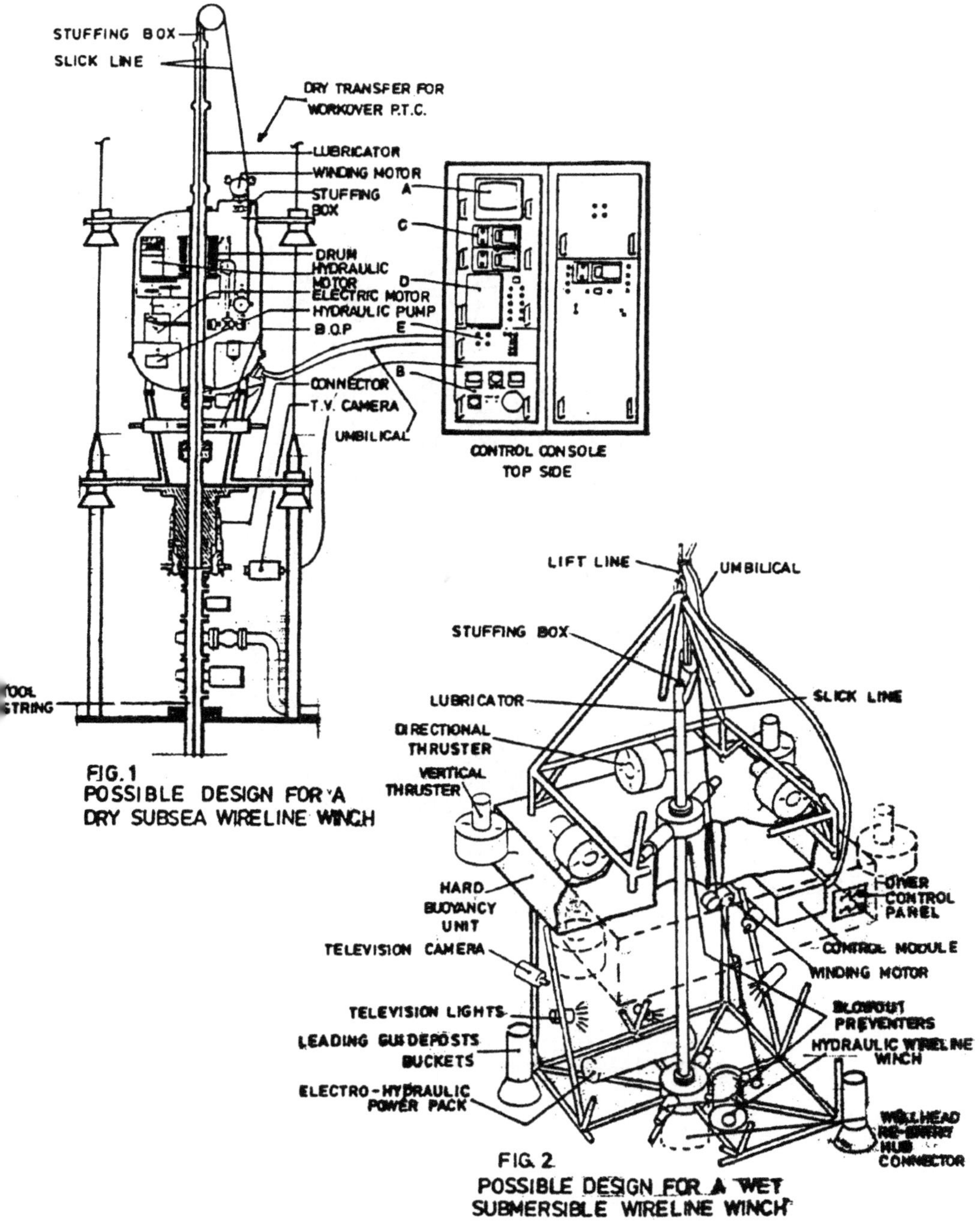

Fig. 23 Workover, subsea wireline winches.

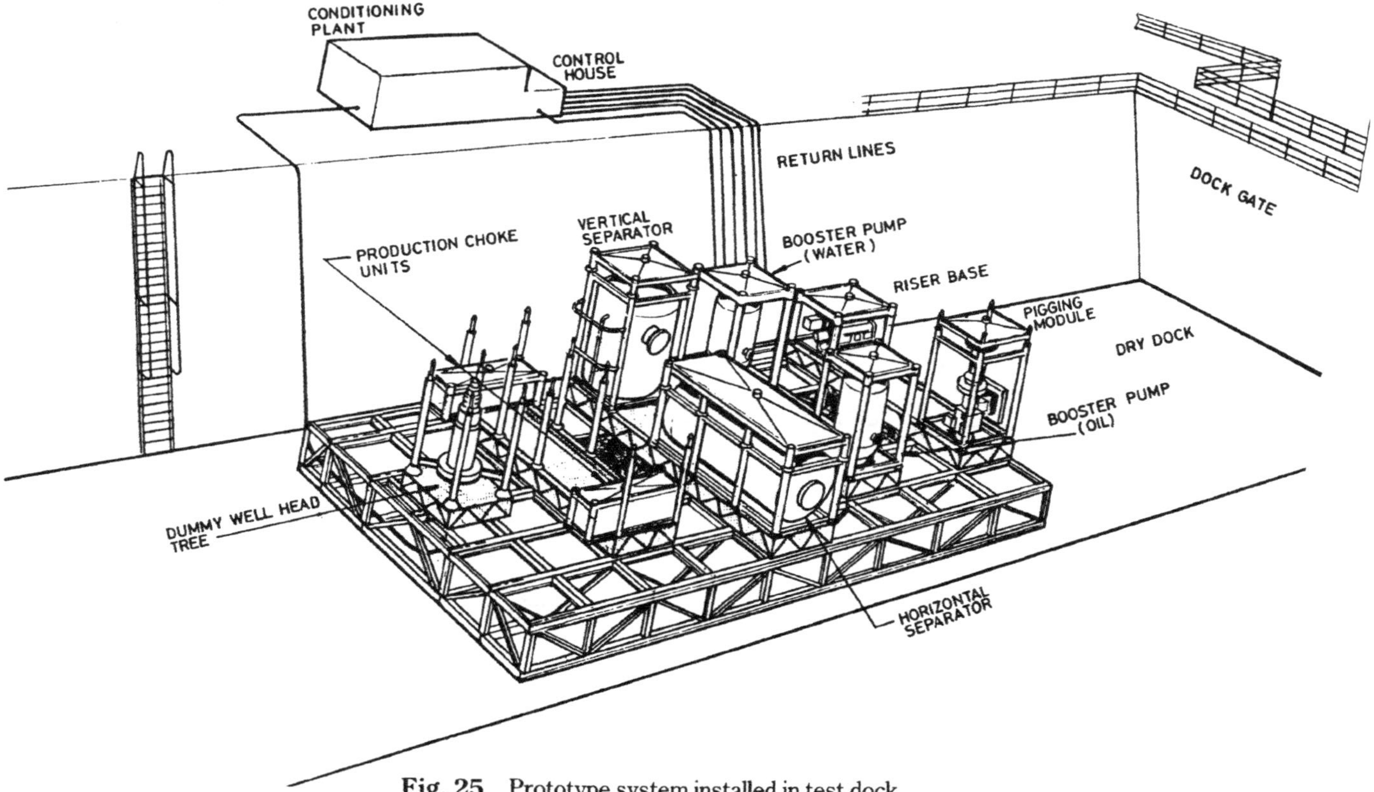

Fig. 25 Prototype system installed in test dock.

Table IV

Distribution of undeveloped discoveries and marginal field development alternatives

	Water depth (ft)						
Region	*1–100*	*101–200*	*201–400*	*401–660*	*661–1150*	*Over 1150*	*Total*
South America	26	23	29	9	5	2	94
Europe	41	55	97	46	13	11	263
Middle East	32	5	11	0	0	0	48
Africa	95	73	73	14	7	2	264
Far East	77	79	108	16	7	6	293
World total*	271	235	318	85	32	21	962

*Except North America, USSR and Eastern Europe.

System	*Field and location*
Steel caisson	Rolf, Denmark
Artificial island	Endicott, US Arctic
Converted jack-up	Espoir, Ivory Coast
Refloatable concrete platform	Schwedeneck, West Germany
Refloatable steel platform	Maureen, UK
Fixed steel platform	Kotter, Netherlands
Fixed wellhead/floating storage	Bohai, China
Subsea completion/floating storage	Tazerka, Tunisia
Converted semi-submersible	Argyll, UK; Green Canyon 29, USA
Purpose-built semi-submersible	Balmoral, UK
Tension-leg platform	Hutton, UK

- underwater power and distribution;
- subsea processing/multiphase flow;
- remote controls and robotics (remote maintenance systems);
- underwater operations and well servicing.

A significant EEC-funded project (see Fig. 25) is now under way and will address these activities, and through reliability/risk analysis, prototype trials and evaluation, will highlight areas for improvement including interconnections, modularization, multiphase flow, installation and retrieval. Subsea systems still have some depths to plunge and distances to cover.

FUTURE TRENDS

With the future trends to smaller fields (less than 100 million barrels recoverable reserves; see Fig. 24) and lower oil prices, the term "marginal" takes on a significant meaning.

The distribution of undeveloped discoveries worldwide, and the alternative options available for field developments with the addition of subsea systems, are shown in Table IV.

In the North Sea a further significance is the distance of satellite field from existing field developments. A target would be to minimize the distances from 10 km to say 30 km over which they could be produced into existing facilities.

For subsea systems to improve in reliability and performance, key areas of the technology will be advanced such as:

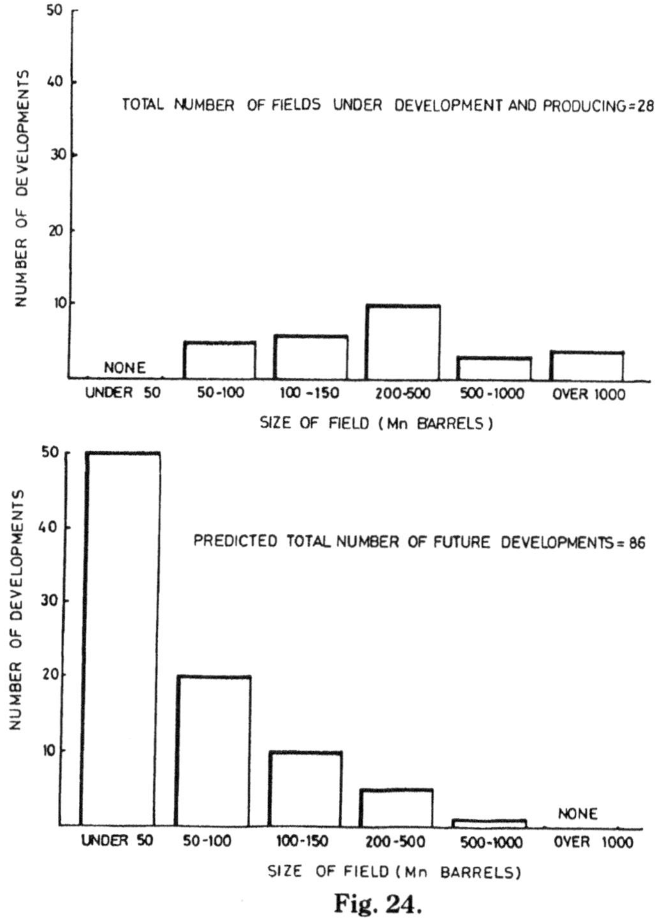

Fig. 24.

ACKNOWLEDGEMENTS

The author would like to acknowledge the support and contribution to the paper from: Conoco, Jack Mercier; BP, Jim Hardy; Hamilton Bros; Shell, Don Henery; Texaco, Brian Walter; and numerous offshore contractors and subsea equipment manufacturers.

Abstracts from the book *Offshore Engineering Development of Small Oilfields*, published by Graham and Trotman.

2

The influence of installation in the design of the Scapa development

G. A. Scott, Occidental Petroleum (Caledonia) Ltd, Aberdeen
M. H. Chancellor, J. P. Kenny & Partners, Aberdeen
and H. van Dalen, Hereema, Aberdeen

INTRODUCTION

The Occidental Consortium's newly developed "Scapa" field has now come on-stream as planned, some 14 months after Department of Energy approval. The Scapa development consists of a major subsea facility connected to the Claymore platform, 3 miles* away by two pipeline bundles and associated control umbilicals (see Fig. 1). The new topside facilities are connected in turn to the pipelines by a large bundle riser, carrying the produced oil and lift gas, with a facility for a utility if required. The field is estimated to have recoverable reserves of 40 million barrels of oil with a gravity of 32.5° API to be extracted from a Lower Cretaceous reservoir.

The Scapa development had a considerable engineering content. The main production riser and template each required a large proportion of this engineering for the installation design. It was a requirement that all of the subsea work should be completed within one season. To accommodate this the contracts were broken down into elements:

- template – fabricate and install;
- bundle – fabricate and install;

Advances in Underwater Technology, Ocean Science and Offshore Engineering, Volume II – Underwater Construction: Development and Potential

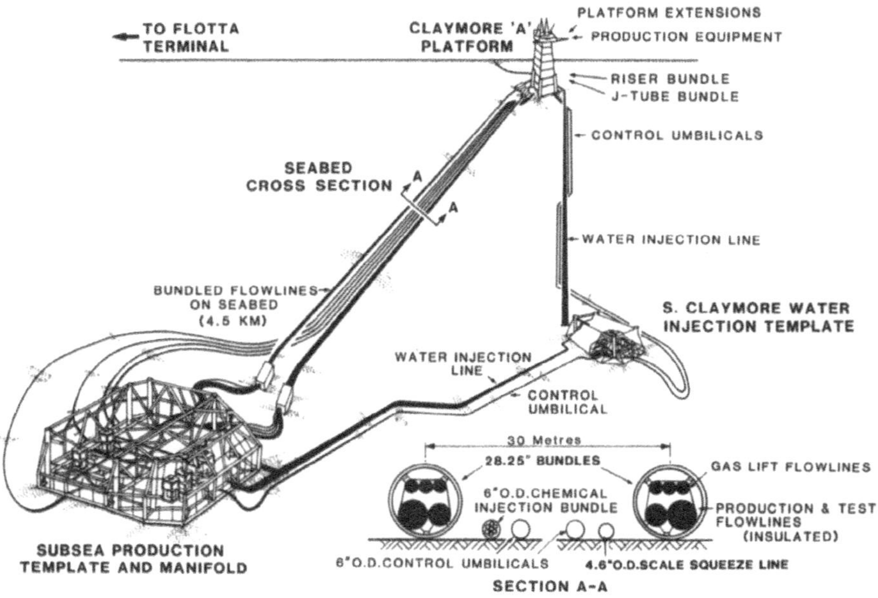

Fig. 1 Scapa subsea development.

- riser – fabricate;
- riser – install.

These packages therefore gave the flexibility to maximize the offshore work programme involving vessel availability and installation capacity. This chapter gives a short overview of Scapa's facilities and installation difficulties, together with a fuller description of the riser and template.

It should be noted with interest that the total weight of Scapa additions to the jacket is approximately 800 tonnes, or 2.3% of Claymore's jacket and topside weight. The structure's capability of accepting the extra load required verification.

FACILITIES

Pipelines and umbilicals

The template is connected to the Claymore platform by an extensive pipeline system, consisting of two 28 in. * multi-line composite bundles, an 8 in. water injection line and a 3 in. bulk chemical treatment line, as shown in Fig. 1.

Each pipeline bundle is approximately 2.8 miles long and contains a

*1 in. ≈ 2.54 cm; 1 mile ≈ 1.6 km.

10 in. production line, a 6 in. test/service line and three 3 in. gas lift lines. The bundles are connected to the template manifold and the platform production riser by flexible spools.

The two pipelines, which are the longest of their type to be towed to date, were installed using the mid-depth tow method. Using the natural buoyancy of the pipeline to raise the line off the sea bed and chains to control this flotation, the lines were towed using two tugs. These tugs kept a set tension on the line during tow and the level of tow was varied depending upon the weather conditions. The towlines were attached to the bundles by specially designed tow heads. These tow heads also served to splay the integral lines for accommodating the flange tie-in points. To confirm manual calculations, tests were carried out on a scale model of the tow head to determine hydrodynamic factors. A full time-domain dynamic computer analysis was undertaken to establish acceptable sea states under which the tow could proceed. This analysis was also used in failure analysis to establish the results of a major component failure, such as in the leading or trailing tow wire.

The two bundles were laid in a corridor approximately 30 m wide. This corridor was also used for the laydown of the two control umbilicals, chemical injection umbilical and scale squeeze line. Each line was sufficiently flexible to be wound on to a carousel on the lay-ship. The messenger line, which was previously installed in the J-tube riser, was transferred over to the lay-ship. Once connected, the umbilical was pulled through the bellmouth up to the surface hang-off point, which was at +68 ft* elevation. The umbilicals were required to be pulled through the shape of an "S" bend in the J-tube riser described in the next section. The resistance to pulling was lowered as much as possible by making the radius of each curve on the swan-neck as large as possible. After locking off at +68 ft the lay-ship then steamed along the corridor, laying the umbilical as it went. Accurate positioning was achieved using a remotely controlled crawling vehicle on the sea bed, which buried the line as it moved.

The water-injection requirements were met by means of a flexible pipeline laid between the South Claymore water-injection template and the Scapa template, 2 miles distant.

J-tube riser

The J-tube riser (see Fig. 2) is a 36 in. diameter caisson with four 12 in. diameter internal pipes which are used as supports and guides for the umbilicals during the pulling operation.

*1 ft ≈ 30.5 cm.

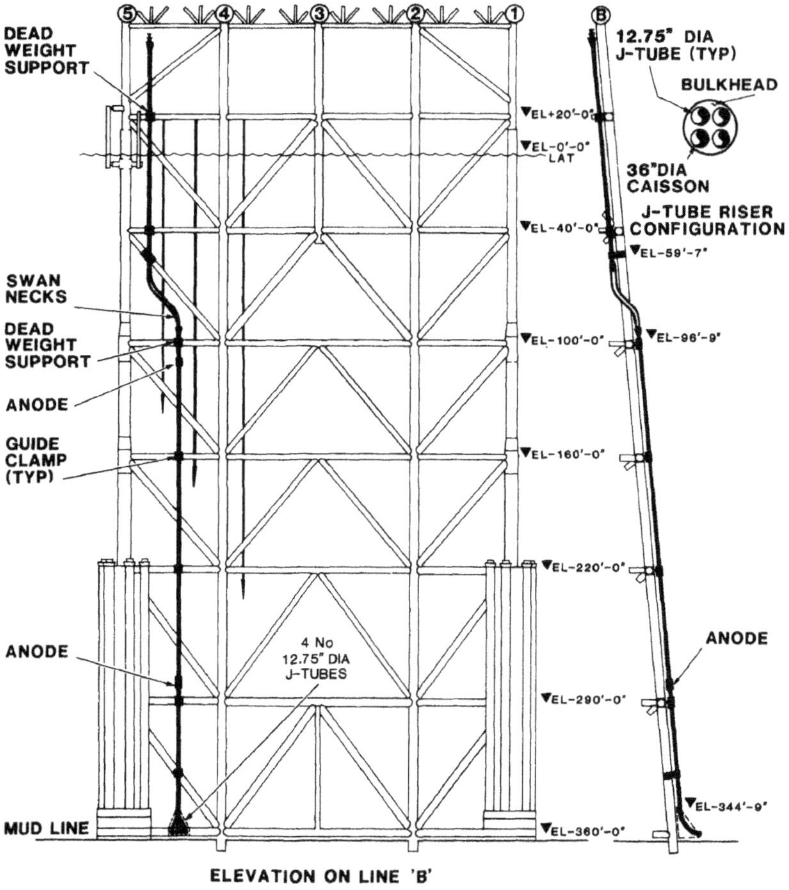

Fig. 2 Claymore (Scapa) general arrangement of J-tube riser.

It was required that the J-tube riser be protected from vessel impact, so a subsea dog-leg was introduced. The riser was therefore installed in three parts, viz.

(a) Section 1 – from sea bed to -100 ft, incorporating four splayed bends, each incorporating a bellmouth;
(b) Section 2 – four large-radius swan-neck spool pieces from -100 ft to -50 ft to ease pulling forces;
(c) Section 3 – top section from -50 ft to +50 ft.

The J-tube riser sections were installed using a combination of the MSV *Tharos*, a semi-submersible vessel with a crane capacity of approximately 300 tonnes, and platform rigging. The tolerances in the clamp-pipe interface were particularly tight and, together with the clamp neoprene liners, caused a certain amount of obstruction to the installation of the lower section. The upper section was difficult to install, due to limited headroom and the proximity of the support trunnion for the main production riser.

Topside facilities

The topside additions are illustrated in Fig. 3.

A 45 ft square extension (Fig. 4), underslung from Module A at the 68 ft elevation was used to accommodate the additional process plant. This plant treats the crude before it joins the main oil-line to Flotta and the gas for enhanced lift. The platform is known as the process pancake and has a structural weight of about 100 tonnes. It was installed using the MSV *Tharos* in one lift, but excluded any equipment or piping. Table I indicates the weights of equipment on the pancake.

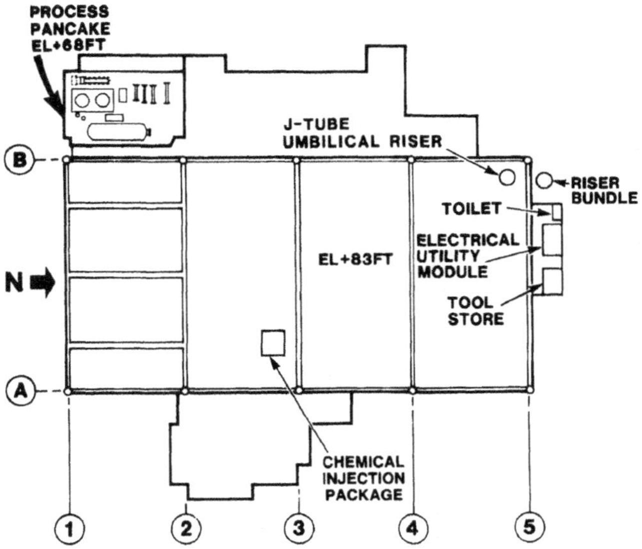

Fig. 3 Claymore platform location plan of Scapa topside facilities.

Module "A", the drilling module, was re-analysed to verify that no overstressed members were present during installation of the pancake or equipment. The results of our analysis showed that all members were within code requirements even after the addition of this large structure.

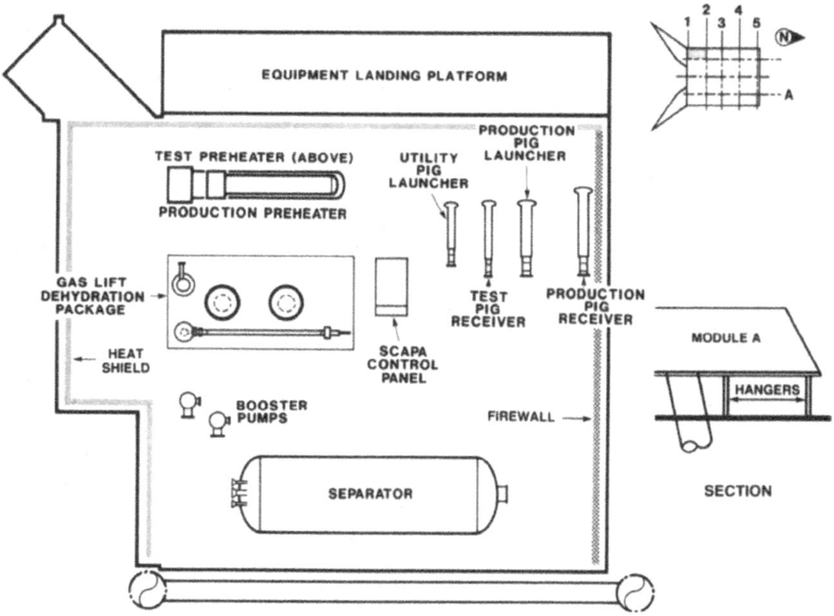

Fig. 4 Scapa pancake.

Table I

Weights of equipment on the pancake

	Weight (tonnes)	*Comments*
Pancake	100	Lift weight during the installation 75 tonnes; final weight includes 25 tonnes of additional steelwork, etc.
Equipment	150	The main gas dehydration vessels (two at 10 tons*) were post-installed
Piping	100	
Services	20	

To cope with the additional electrical and monitoring requirement of Scapa a new utilities module, 36 × 10 × 11.5 ft and weighing 25 tonnes, was to be installed on the North face of Claymore, on the side of Module D, the power-generation module. The location where this new module was to be installed needed to be modified to take the new support structure and to remove an existing tool store. As part of the modifications new tool-storage facilities and an additional toilet were also incorporated.

*1 ton ≈ 1.016 tonne.

The installation of the items on the North face required an in-depth installation study for using the MSV *Tharos*. It was inadvisable to moor MSV *Tharos* alongside the North face of Claymore, because of the close proximity of the main oil and gas risers and the turbine exhausts from Claymore. The scheme eventually adopted required that the MSV *Tharos* be moored at the North-West corner of Claymore, and placed the modules on the helicopter landing deck to attach platform rigging, being then lifted down into place using a combination which was successfully deployed despite the tight tolerances for the placing of units. This is illustrated in Fig. 5.

Also installed underdeck on the 68 ft elevation was a package for monitoring, controlling and pumping the various chemicals required for the chemical injection on the template.

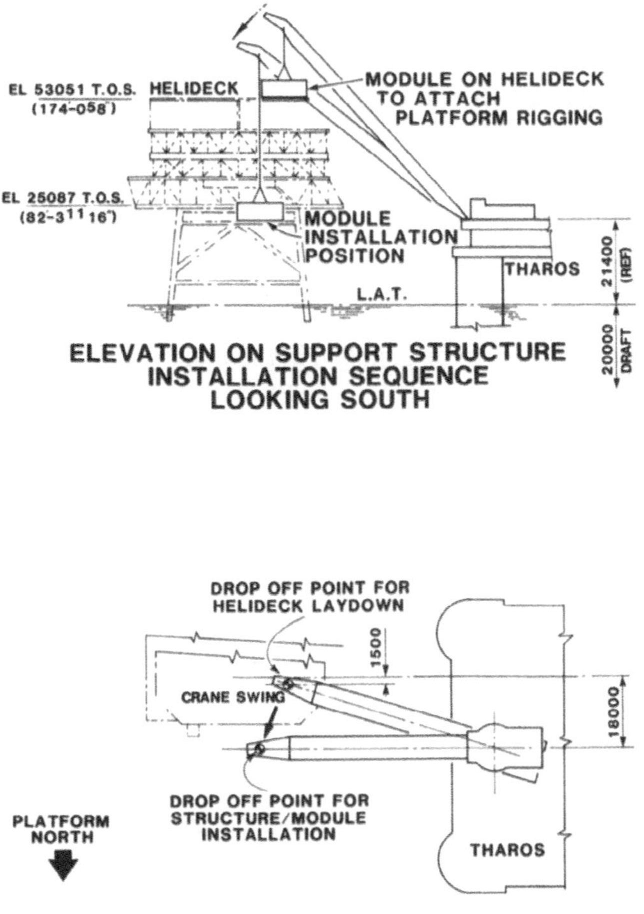

Fig. 5 Elevation on support structure installation sequence looking south and plan view on Tharos location for support structure/module installation.

The chemical-injection package had a very complicated installation scheme because of its location near the centre of the platform. Several cross-hauling operations were required before it was in its desired location.

The weights of the extra additions are shown in Table II.

Table II
Weights of extra additions

	Weight (tonnes)
Utilities platform (steelwork)	10
Utilities module	25
Tool store	15
Toilet block	2
Chemical-injection package	18

PRODUCTION RISER

The main production riser is of bundled caisson construction (see Fig. 6) with outer 36 in. diameter containing 14 pipes in total, two 10 in. production, two 6 in. test utility, two 6 in. spare, six 3 in. gas lift and two 3 in. spare. This riser, at 420 ft and 230 tonnes, is the longest and heaviest to be installed in the North Sea, and presented a difficult and unique challenge. Detailed engineering into various methods of installation led to the use of a strongback attached to the riser, allowing lifting and upending in one continuous operation. The successful installation of the riser by this method relied upon the capability of the dual-crane SSCV *Hermod* to offer precise control during the handling of the riser/strongback assembly throughout the lifting/upending/transfer operation. A double-drum winch located on the 68 ft level of Claymore was used to lower the riser into its final location. The rigging for this operation was supported by a special "heavy-duty" lift beam (Fig. 12, below) attached to a strong point on the jacket.

Historical

In 1981 Occidental installed a riser of similar diameter on the North-East corner of the Claymore jacket to service the North Claymore subsea water-injection facility. This riser was 50 ft shorter and contained less internal pipes than the Scapa riser. The method of installation used for this previous riser utilized a series of davits from the side of a heavy lift vessel.

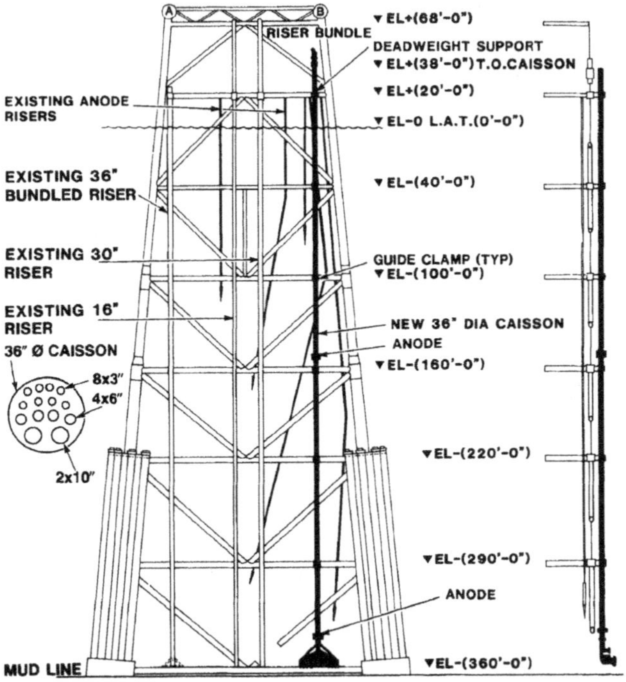

Fig. 6 Claymore (Scapa) general arrangement of main production riser.

Constraints

There were several criteria which required fulfilling which had a constraining effect:

- the height available for installation;
- the riser was made as long as possible, to reduce offshore hook-ups;
- sea-bed clearance was left for divers.

Options

The method of installation used was primarily dictated by the difficulties in safely upending the riser without overstressing the caisson. This was due to its length and weight. The available headroom was of particular concern when transferring the riser to the platform rigging.

Initial studies into conventional upending methods using dual-crane lifts showed that the riser would be overstressed during upending if it were unsupported along its length. Two alternative methods were investigated. First, the use of davits, as previously used, hung over the side of the SSCV. Secondly, the use of a strongback to support the riser during upending.

The davit system consists of four davits spaced along the riser, with the crane attached to the top end of the riser. The davit wires were to be systematically lowered to upend the riser gradually over the side of the SSCV.

The strongback method (see Fig. 7), uses an 84 in. pipe clamped to the riser, with trunnions on the strongback top and bottom. The arrangement is lifted horizontally off the cargo barge and upended using the two SSCV cranes. The strongback system has a number of advantages over the davit system, and was adopted for installation.

The main advantages of the strongback system is the more precise handling control during lifting and upending, which can be performed in one operation over a short period. Using the davit system it was

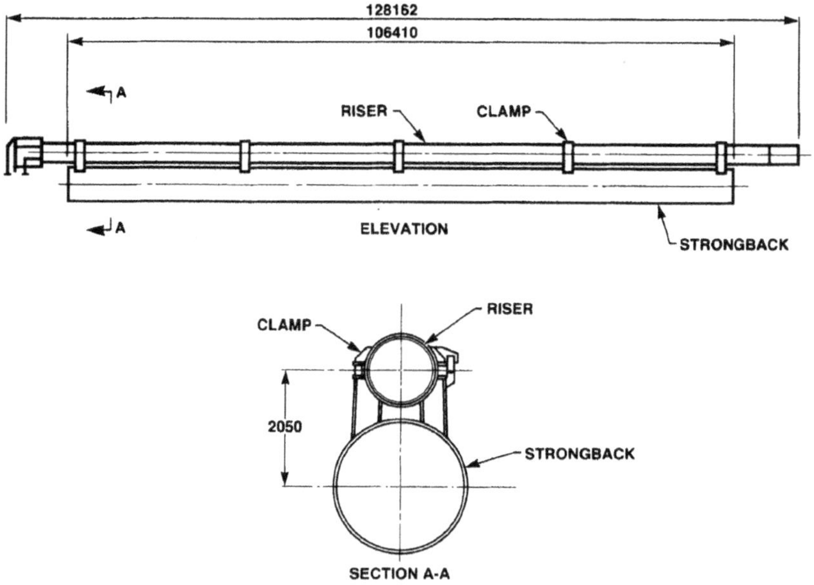

Fig. 7.

calculated that some 250 steps over 2 days would be required, considerably increasing the risk of the weather affecting the installation and the consequences of delaying the whole operation.

Other advantages were the relatively straightforward sea-fastening and transportation of the riser/strongback assembly on a cargo barge, as against the need to transfer the riser to the SSCV in sheltered water for securing to the davits before transportation to site. It was also found that when upending using davits the riser would interfere with the SSCV anchor wires, with a time delay to slacken off the wires and an increased risk of damage to the riser due to the duration and weather-sensitivity of the operation, and the proximity of the riser to the side of the crane vessel.

Method

General

The riser was supported for transportation and upending on an 84 in. strongback. After upending using two cranes and supporting from the stern of the SSCV, the five riser support clamps were opened and the riser handed over to the pre-installed platform lifting equipment for final positioning into the platform clamps. Installation was completed by closing the platform clamps.

Detailed procedure (see Figs 8–10)

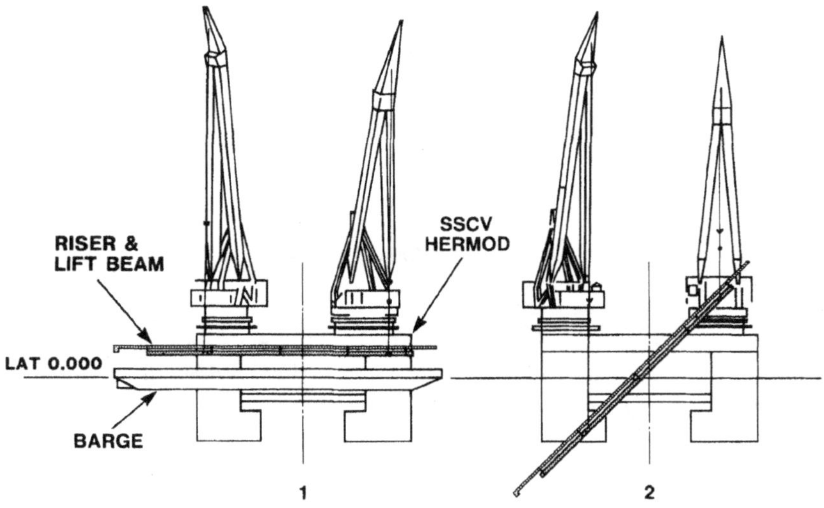

Fig. 8 Riser lift from barge and upending.

1. Riser/strongback assembly was transported to site on a cargo barge pre-rigged for horizontal lift, upending and hang-off from the SSCV.
2. During the horizontal lift-off from cargo barge and upending (dual-crane operation) each crane was attached to the lifting trunnions which were provided at top and bottom of the strongback. At the end of this sequence the entire weight of the

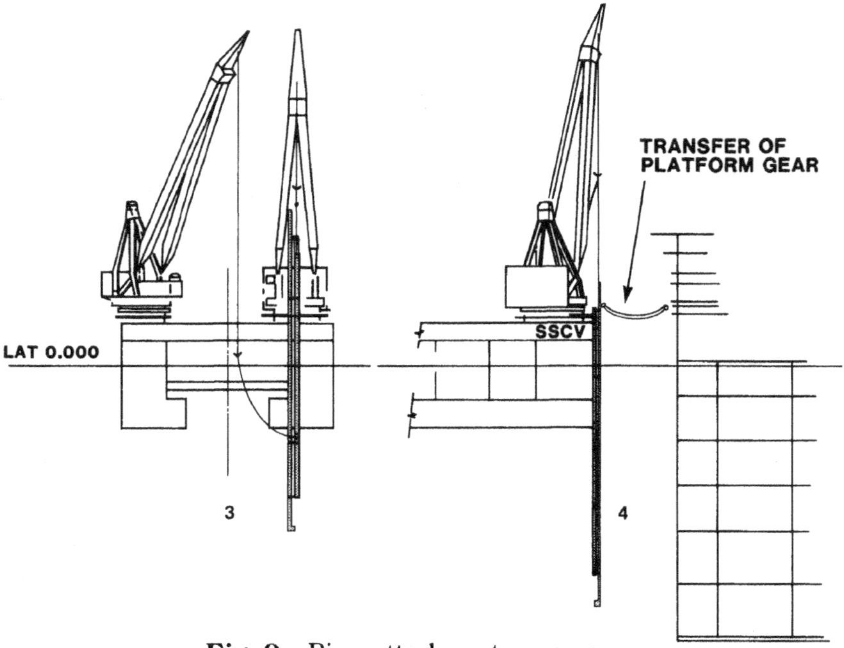

Fig. 9 Riser attachment sequence.

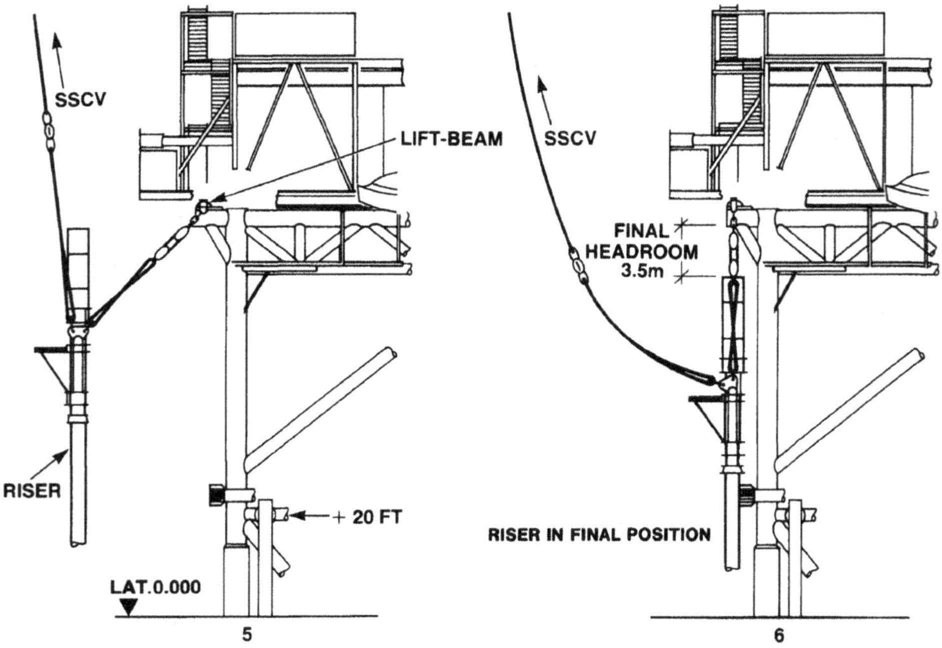

Fig. 10 Elevation on platform (looking east) for riser installation sequence.

assembly is taken by one crane (sequence 1–3).

3. Lower slings were released by divers. The lower trunnions incorporated a specially designed hinged keeper plate detail to facilitate removal.
4. Released lower clamps were released and upper riser clamps on the strongback were partially released.
5. Assembly hung off stern of SSCV using pre-rigged slings on the upper trunnion. The upper trunnion was designed for a dual function, containing both lifting and hang-off slings (sequence 4).
6. Attach platform and SSCV handover rigging to riser. Handover was achieved using a triangle plate to which both sets of upper rigging were attached. The lower rigging was common to both platform and SSCV arrangements. During the handover operation (sequence 5) the triangle plate rotated, gradually transferring load from SSCV rigging to the platform rigging. The two sets of platform rigging were reeved through 12 part blocks, and as such the transfer was necessarily slow. This also provided very fine control during installation into clamps.
7. Positioning of riser in jacket clamps was achieved using installation guides at two elevations on the jacket, with positioning tugger lines from the riser to the jacket available if required.
8. When the riser was fully supported by the platform rigging (sequence 6 and Fig. 11) the SSCV rigging was released. The riser was lowered to its final elevation and jacket clamps closed by divers.

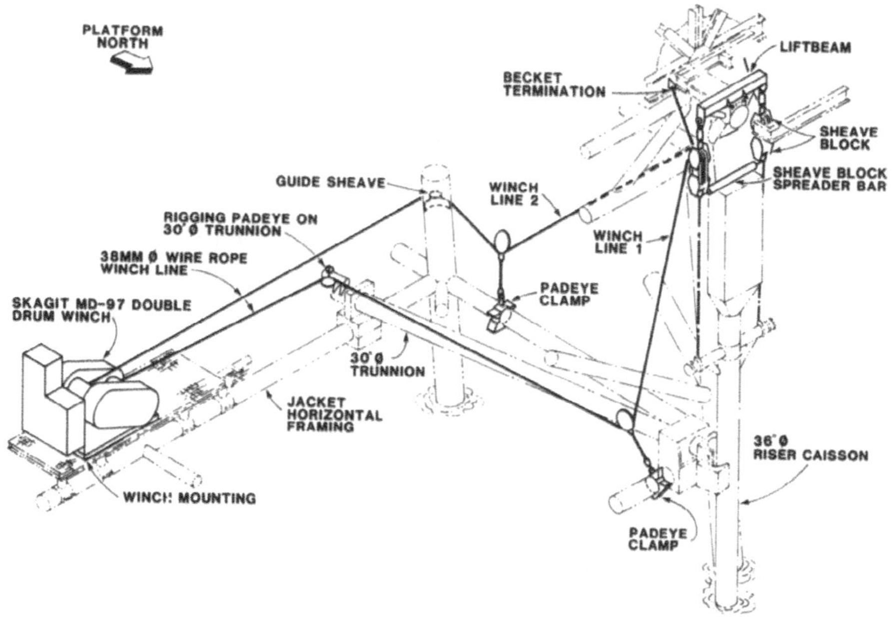

Fig. 11 Platform caisson installation scheme.

Discussion

The installation scheme as described illustrates the detail to which design engineering was applied to the installation of the main riser. Several areas were a compromise of other disciplines' requirements, e.g. the top of the riser was required to be as high as possible to reduce offshore hook-up time and to take the individual pipes out of the splash zone as far as possible, yet enough headroom had to be maintained to accommodate the rigging and sheave blocks. Because of the shallow headroom available close headroom blocks were specified. As can be seen from Fig. 11, the route taken by the rigging from winch to lift beam (Fig. 12) was tortuous. Each position required careful checking for load-carrying capability as well as for site fouls.

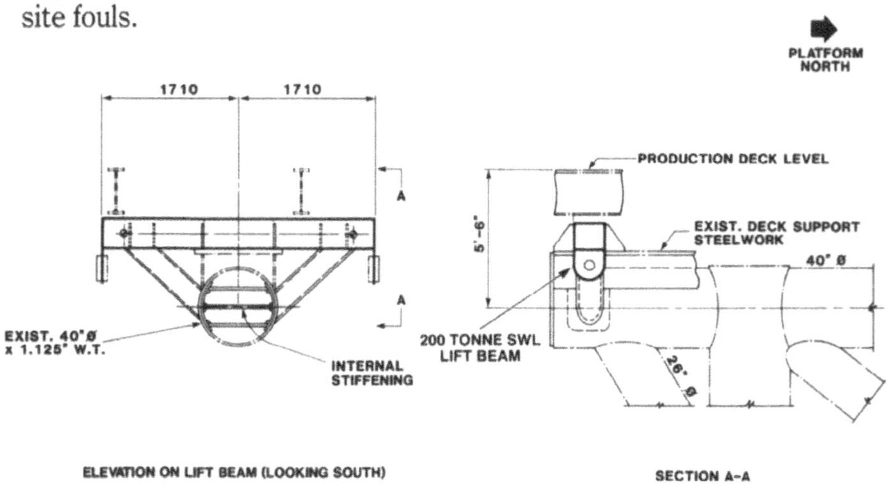

Fig. 12 Elevation on beam (looking south) for riser caisson installation.

TEMPLATE

The Scapa production template (see Figs 13 and 14) and manifold provide a facility capable of accommodating a maximum subsea design production rate of 33 000 barrels/day of fluid. The template structure includes the following facilities:

- drilling support;
- Christmas tree support;
- process piping (production, test, gas lift and water injection);
- controls and instrumentation;
- chemical injection;
- tie-back of off-template wells;
- protection from external damage and from hydrodynamic loading.

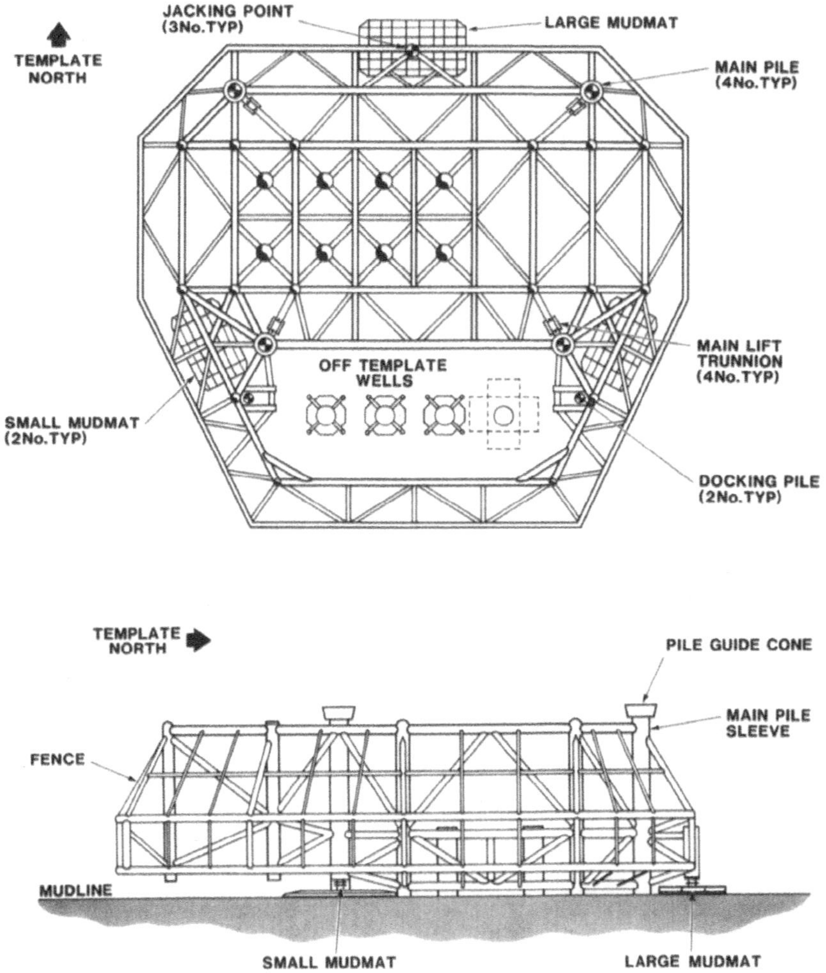

Fig. 14 Section of template.

The resulting template structure weighed 920 tonnes in air, and had an approximate in-place operating weight of 400 tonnes. The template was designed to accommodate eight on-template wells and provided an enclosing protection fence and tie-back for four off-template wells. Three of the off-template wells were existing at the time of installation.

In addition to the above production criteria, the template was designed to be installed as a single item by a four-point lift from a semi-submersible crane vessel. It was required to be accurately positioned over the existing wellheads and fixed to the sea bed within the following tolerances:

- positioning within +-0.5 m of design location;
- orientation within ± 1° of existing well orientation;
- level within ± 0.3° of the horizontal.

Accurate positioning of the template structure, relative to the three existing wells, was achieved using two docking piles. These docking piles were placed using a temporary docking pile frame, tied back to temporary protection frames on the two extreme existing wellheads.

The levelling system used consisted of three mudmats operated by individual hydraulic cylinders and controlled by a diver-operated console, centrally located on the template. After final levelling the template was secured to the sea bed by four corner piles, each pile fixed to the template structure by swaging.

Operational requirements

Drilling support

The primary drilling support requirement was that the eight on-template drilling slots should limit tolerance to +0.3 degrees from the vertical. This was achieved through strict control during fabrication and in the use of the in-built mudmat levelling system.

In addition, the wellbay framing was required to provide support for drilling, workover and production equipment.

Christmas tree support

The required Christmas tree support arrangement centred upon the provision of a wellhead housing receptacle for landing and running casings, incorporation of the tubing hanger, support of the permanent guide base, and tie-back of production, gas-lift and water-injection pipework.

Process piping

The template structure was designed to provide area and support for a skid-mounted process piping manifold. This skid was installed on the template before installation, but was required to be capable of being lifted out of the structure as a single unit to allow subsea replacement, if operationally necessary.

Controls and instrumentation

The template is provided with manifold hydraulic interface panels which provide regulated pilot signals to control valves within retrievable subsea control pods, located on the trees and manifold pipework of the templates. The pilot signals permit hydraulic fluid to be directed to the appropriate hydraulic operated valve (HOV) and thus operate the designated valves.

Chemical injection

The template is provided with a chemical injection manifold, on the western side of the template, which must provide for the injection of six different chemicals to the six designated producing slots.

External loading

As the template is a registered installation, it is charted and will be buoyed after completion of drilling.

The template is required to offer mechanical protection against the impact of trawl boards and medium-sized "hooked" anchors. To achieve this, a sloped outer fence was provided, extending completely around the template wellslots and manifold, and the off-template wells, rising above the height of the internal Christmas trees. Absorption of impact energy is achieved through the local plastic deformation of the secondary fence members.

The template is additionally required to remain stable under the action of hydrodynamic loading, up to and including the 50-year storm condition.

Installation requirements

There were two primary requirements for the installation phase of the template. These requirements were: that the template be installed safely over the existing off-template wellheads, with adequate provision made against damage to the wellheads during this operation; and that the installation method should ensure the accurate positioning of the template relative to the wellheads, within the tolerance limit specified.

The method used to achieve the installation requirements is discussed in the following section.

Method

Design and fabrication

The design and fabrication phases of the template structure were required to be completed within a tight timespan in order that installation of the template could progress on schedule. Priority was therefore given to finalizing the conceptual design of the template and establishing preliminary material take-offs so that material orders could be placed before detailed design of the template began.

The detailed design of the various template components was structured to comply with fabrication requirements, priority being given to those sections requiring the earliest fabrication. In this way the fabrication of the template could be progressed during the detailed design phase, the template design allowing the simultaneous fabrication of different component items, to be pieced together at a subsequent date. In particular, the whole of the production manifold piping and pigging loops, including cable trays, hydraulic control panels and control pods to HOVs, was constructed as a separate skid assembly and installed on the main template before load-out. The skid system allows subsequent retrieval from the template as a single unit for maintenance and repair of equipment.

Loadout and lifting

Transportation to site was undertaken using a flat-topped barge, the template having been skidded out on to the barge directly from the fabrication site.

The use of a four-point lift may induce a racking effect and high torsional loads in the structure members, due to the sling configuration and potential uneven sling distribution. In order to eliminate lifting frames, sling protection frames were incorporated to utilize the four-sling arrangement. The structural design had to reflect, therefore, this requirement for a stiff, torsionally-resistant structure in addition to one capable of resisting the dynamic amplification effects and the air/water interface loading normally associated with the installation of a subsea structure.

Precise weight monitoring to determine the centre of gravity and centre of buoyancy locations, careful rigging design and selection of the lift points were required to ensure that the template remained level during the lowering operation, to avoid the possibility of hang-up on the docking piles. Additionally, to lessen the risk of hang-up, bucket-type guides were used on the template.

The obvious location for the lift points was on the four 44 in.

vertical tubulars forming the main pile sleeves. These tubulars form the central columns of the template structure and are, therefore, natural strong points. In addition, they are located away from the template's centre of gravity at a sufficient distance to provide a stable lift configuration.

Before installation a sea-bed survey was carried out using a diving support vessel, and the sea bed was levelled sufficiently to ensure that the level requirement could be accommodated within the stroke of the hydraulic jacks.

As for the main riser installation, specially designed lifting trunnions incorporating a hinged keeper plate were used to facilitate diver removal of the template lifting slings.

To avoid damage during sling removal, extensive temporary equipment/pipework protection was required. The protection was designed using a minimum number of bolted connections and incorporating guidance pins to ensure quick but safe removal by divers and lifting.

Docking piles and guide frames

The function of the docking pile guide frame (Fig. 15) was to ensure accurate location and guidance for the two template docking piles.

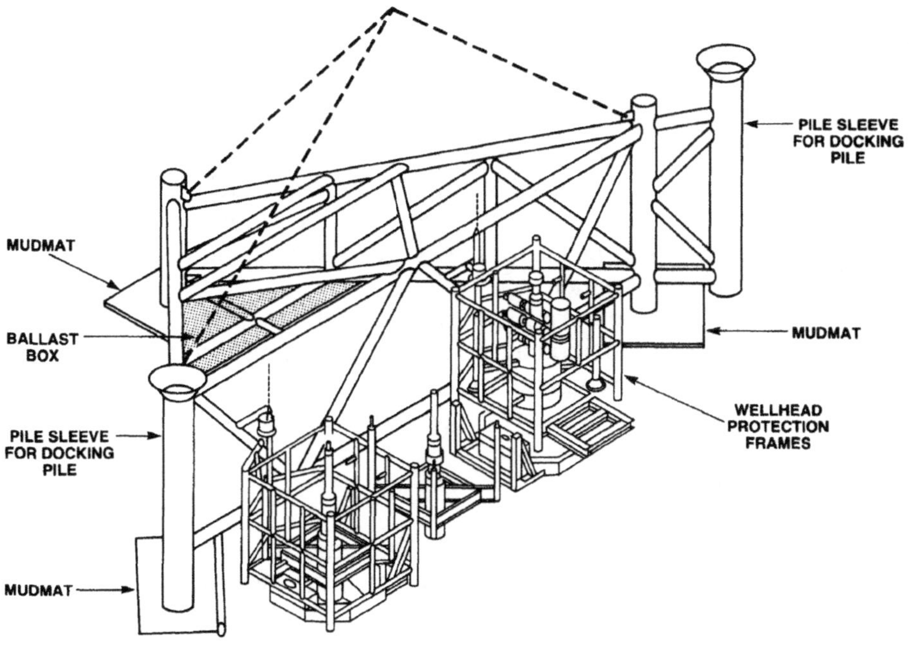

Fig. 15.

The frame consisted of a triangular stiffened truss arrangement, constructed of tubular steel and with the pile sleeves located at the two extreme points of the triangle base.

Stability of the frame to sustain pile-driving loads was achieved by flooding the majority of the frame members, thus providing extra ballast to the frame. Stabbing cones on the frame were provided to interlock with the guide pins on the wellhead protection covers. On-bottom support for the guide frame was provided by mudmats at each corner of the triangular structure.

Tubular braced frames were bolted to the guide pins on the permanent guide bases to provide initial locating and alignment for the docking pile guide frame. They also offered protection, during this operation, to the tree and equipment on the well head.

Installation of the frame was by three-point lift, with taut wires feeding themselves through the stabbing cones on the guide frame, and attached subsea to the guide pins on the wellhead protection frames.

Before the initiation of pile-driving operations the clamps securing the wellhead protection frames were released to minimize shock-load transfer to the wellheads. Both guide piles were driven to a penetration depth of 25 m; however, one pile was left 2 m higher than the other to allow sequential location of each pile. The pile heights were sufficiently large to ensure that during the subsequent template installation (Fig. 16) lock-on to the piles was achieved before encroachment below the height of the existing water-injection tree.

When pile-driving operations were completed, retrieval of the frame was accomplished first by cutting the frame locally at both docking pile sleeve positions and recovering the frame with a three-point lift, and subsequently recovering each pile sleeve. Retrieval of the frame and the pile sleeves in this manner was intended to minimize possible damage to the docking piles and the wellhead protection frames.

Throughout the template-lowering operations (Fig. 17) and, in particular, for final positioning just before docking, the positions of the guide frame and template were monitored by divers, ROVs and side-scan sonar, ensuring no damage to the existing wellheads.

Levelling system

Once installed subsea, the levelling of the template was accomplished using the three-point hydraulic jack and mudmat system built into the template structure. The system was operated from a centrally-located diver control console and was felt to offer the most simple,

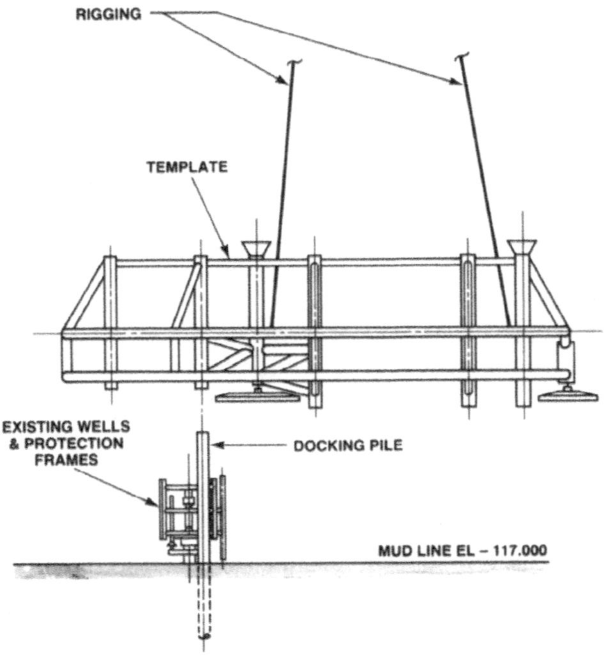
RIGGING
TEMPLATE
EXISTING WELLS
& PROTECTION
FRAMES
DOCKING PILE
MUD LINE EL – 117.000

Fig. 16.

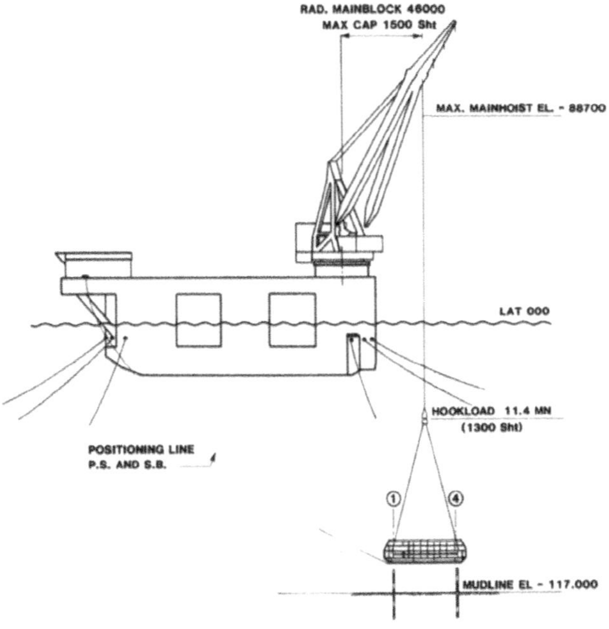
RAD. MAINBLOCK 46000
MAX CAP 1500 Sht
MAX. MAINHOIST EL. - 88700
LAT 000
HOOKLOAD 11.4 MN
(1300 Sht)
POSITIONING LINE
P.S. AND S.B.
1
4
MUDLINE EL - 117.000

Fig. 17.

efficient and cost-effective means of levelling the template.

The jacking system was designed to support the structure's "in-air" weight to enable testing and operation of the system to be carried out in the fabrication yard before load-out. Each jack had a 1 m stroke and 400 tonnes capacity, and was capable of an extension rate of 200 mm/min.

A primary design criteria for the subsea levelling condition was to minimize the template submerged weight, and hence minimize the mudmat bearing areas. To this end, much of the template structure consists of sealed tubular members to increase the buoyancy loads acting on the template. Each mudmat is linked to its respective hydraulic jack by a clevis-pin arrangement. This ensures smooth load transfer to each mudmat, regardless of the inflexibility of the foundation.

The levelling operation was carried out in two stages. Having checked that the template was on a level attitude, the piling of the template was initiated. On completion of the piling the levelling system was used to obtain a final level configuration, accurate to within 0.3° to the horizontal.

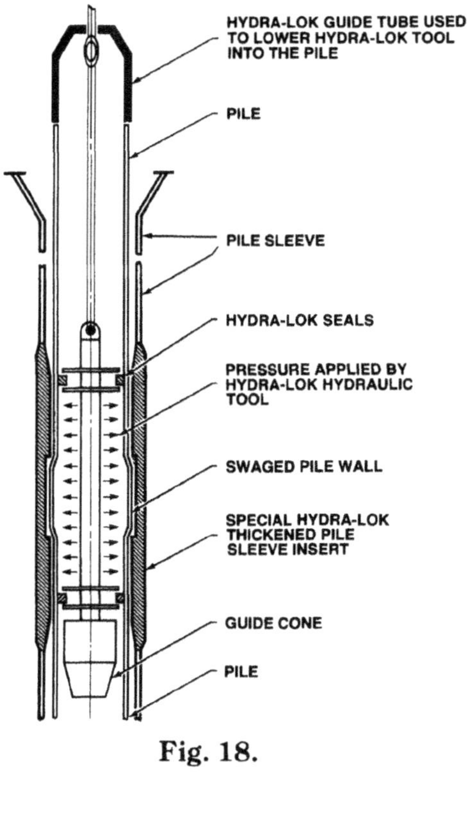

Fig. 18.

Pile and pile fixing

After completion of the initial template levelling and in order to ensure the stability of the template on the sea bed, the template structure was restrained by four 36 in. diameter piles, driven through the 44 in. diameter pile sleeves, integrated within the main template structure. The pile-driving was accomplished using subsea hammers. These pile sleeves are symmetrically located about the template's centre of gravity. The minimum recommended pile penetration was 44 m.

After final levelling of the template, the pile to template fixing was achieved using the BUE Hydra-lok system (see Fig. 18). This system swaged the piles into specially fabricated grooved cans, which made up a section of the pile sleeve between 5 and 6 m above the mudline. A tool was located into the pile tubular at this point, and was used to pressurize this section of pile, causing local deformation into the grooves provided, thus fixing each pile to its relevant pile sleeve. It now remained to remove the guide cones and trim and cap the piles.

From the template arriving, on the barge, at the side of the SSCV *Hermod* to the SSCV leaving the field was 5 days – a commendable achievement for all those involved.

3

The design and installation of multiwell subsea template structures

S. M. Komaromy, J. P. Kearney and M. A. Knight,
Earl and Wright Ltd*
and W. G. Edwards, Humphreys & Glasgow Ltd

INTRODUCTION

For a conventional multiwell offshore field development, traditionally drilling could only commence after a fixed platform had been installed and commissioned. This meant that the gas or oil was not recovered until well into the development programme, and that the cost of developing fields in deepwater or marginal fields became, if not impractical, then certainly less financially attractive.

It is usually cost-effective for an operator to begin production at the earliest opportunity, because recovery of capital outlay cannot commence until the first gas or oil flows. Earlier production can be achieved by pre-drilling the wells through a drilling template previously installed on the sea bed. The wells can then be tied back to surface production facilities typically installed on a fixed platform or a tension-leg platform. The early production of gas or oil must, of course, be offset against the costs of the template, pre-drilling and tie-back operations.

Where the cost of providing a fixed structure proves to be uneconomic, then the use of a template containing the production

Advances in Underwater Technology, Ocean Science and Offshore Engineering, Volume 11 – Underwater Construction: Development and Potential

* An H & G Company.

facilities should be considered. It can provide the drilling facilities in addition to accommodating Christmas trees, production piping, well killing and pigging facilities. The template can act as a gathering point for satellite wells, and export by pipeline to either a jacket located in shallower water or via a riser to a floating structure within the development. Its use in this fashion allows the economic development of a marginal field and even small pockets of a large field not accessible by deviation drilling from a central platform. The technology used and the expertise gained from such operations will also prove useful in exploiting the reserves in the deepwater fields which are currently being investigated.

The major considerations governing the design and installation of these structures form the basis of this chapter.

DESCRIPTION

General

When a fixed or tension-leg platform is to be used in conjunction with pre-drilled wells it is first necessary to drill some or all of the wells through a previously installed template. The platform is subsequently installed, the subsea wellheads are tied back to the production trees on the deck, and any further wells can be drilled from the platform. Such a template structure, which is only required during the drilling operation, is generally referred to as a drilling template, a typical example of which is shown in Fig. 1.

As an alternative, it is possible to locate the production trees on a subsea template and then tie-in to a floating production facility (e.g. a modified tanker or drilling vessel) or a remote platform. The oil or gas is exported to the floater via a riser or to the platform by flowline. Such a structure is commonly referred to as a production template (a typical example is given in Fig. 2), and is expected to act as a guide and support for the drilling operations and to house the Christmas trees throughout the life of the installation.

For both groups there are two methods of manufacture: modular or unitized. The modular method consists of a series of interlocking single-well modules and may be used for small developments where the exact number of wells has not been established. The size of the modules means that they are suitable for installation by a drilling rig, and they can be added to the main group of modules as required. The unitized system concerns larger single-piece structures from one well slot, up to 40 slots in some cases.

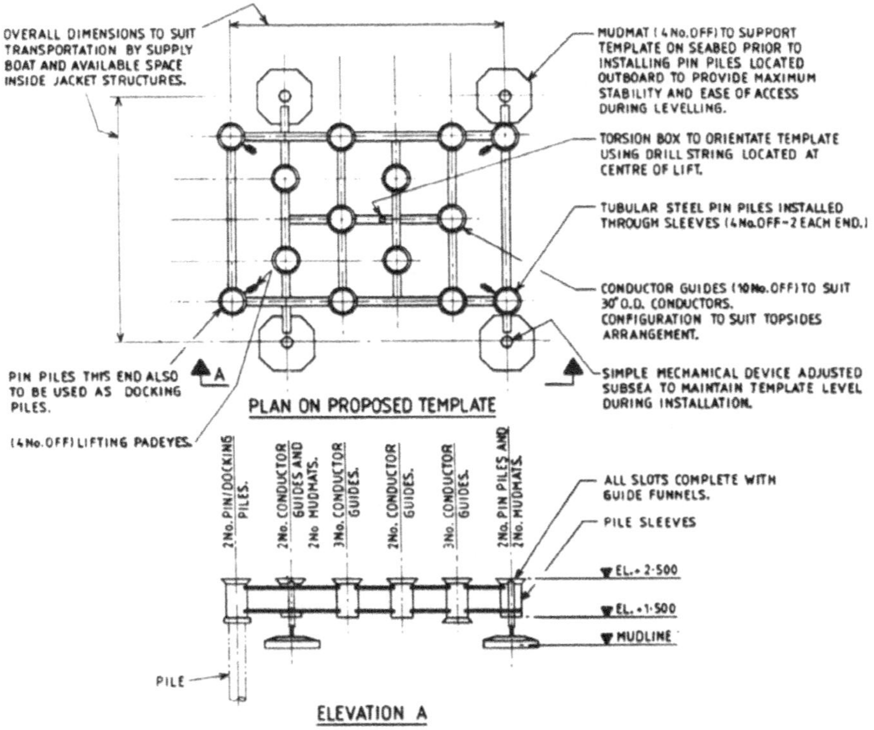

Fig. 1 Typical subsea drilling template.

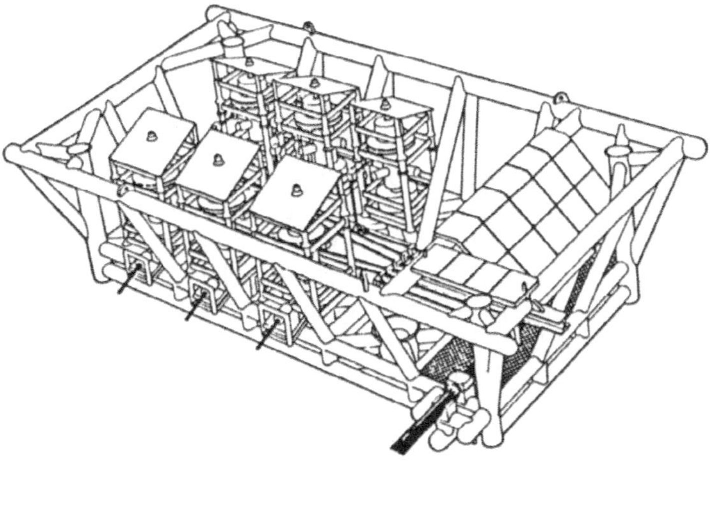

Fig. 2 Typical subsea production template.

This chapter considers only the design and installation of the unitized type of template.

Drilling templates

Drilling templates are small, relatively simple structures which are only required during the drilling phase of operations. A series of wells can be drilled through the template from any appropriate drilling vessel, such as a semi-submersible drilling vessel or a jack-up. If a jacket is to be installed over the template, then it is located precisely using an arrangement of bumpers or docking piles which are fixed relative to the template. Since the wells are pre-drilled, completion or "tie-back" can take place as soon as hook-up and commissioning of the topsides is complete.

In its simplest form the drilling template may be a single-level plate girder or tubular structure supporting wellhead receptacles on a grid which is compatible with standard drilling equipment, typically 8 ft 6 in.* between wells. The depth of framework surrounding the receptacles should be sufficient to provide strength and rigidity for the structure and good support at a suitable elevation for the wellhead receptacle.

The template will usually be temporarily supported and levelled off mudmats which are connected to the structure via hydraulic rams. To ensure that there is a positive location for the first well to be drilled and that there are no restrictions on the order of drilling the array of wells, it is usual to provide permanent support to the template by piling, which is either driven or drilled and grouted into place. This also prevents problems occurring due to soil washout during well-drilling operations. The template is then attached to the piles by grouted sleeves or by swaging.

An example of a drilling template installed in the UK sector of the North Sea is given in Fig. 3. The template consists of an arrangement of receptacles spaced exactly in line with the conductor guide framing of the main jacket structure. The receptacles are supported by a tubular steel grillage which transmits the applied loads to the piles.

In order to produce from a drilling template it is necessary to tie-back the wellheads to the production facility. The term "tie-back" is derived from the fixing or coupling together of the casings, and is used to describe the overall operation of completing the casings from the wellhead up to the drill floor. The general arrangement of such a system is shown in Fig. 4. The tie-back conductor is usually a 20 in. diameter tubular lowered to stab either over or into the wellhead,

*1 ft ≈ 30.5 cm; 1 in. ≈ 2.54 cm.

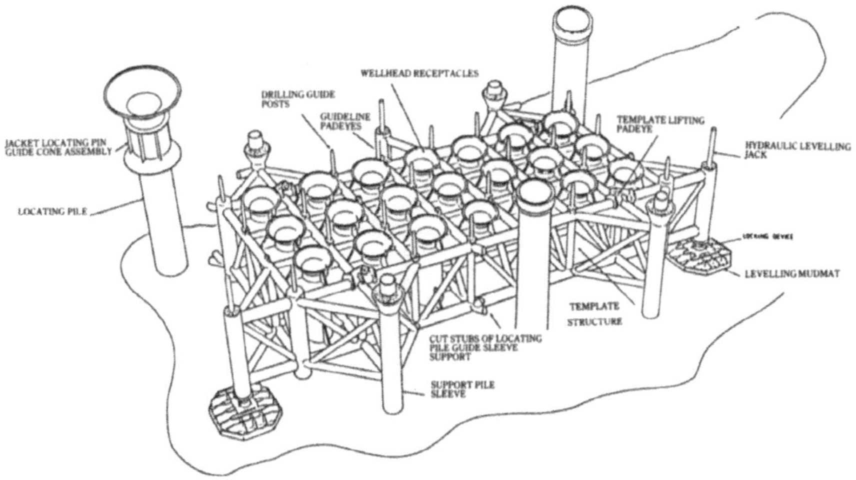

Fig. 3 Isometric drawing of drilling template.

and is sealed to the wellhead casing. To ensure a trouble-free tie-back operation, great emphasis is placed on controlling the verticality of the wellhead and minimizing the lateral offset of the jacket relative to the template. The maximum amount of angular misalignment that can be tolerated is primarily a function of the tie-back equipment and the conductor-framing arrangement in the jacket. The tie-back conductor is designed to be sufficiently flexible to accommodate these misalignments.

Production templates

A production template may be defined as one which is designed to house and protect a combination of drilling, production and control equipment. It must in the first instance act in the same manner as a drilling template, providing both a level base for the wellhead and guidance to the drilling equipment. It must also provide access for pulling-in and connecting the flowlines and service lines. Its most important difference from a drilling template is the requirement to support the production equipment throughout the life of the facility. This requirement influences not only the size of the template, but also the quality control of the fabrication.

It will typically consist of a base structure similar to that of the drilling template, but consideration should be given to providing a trussed arrangement which has the advantage of allowing greater freedom in the positioning of production piping. Above the base, at an elevation at least level with the top of the equipment, a ring of

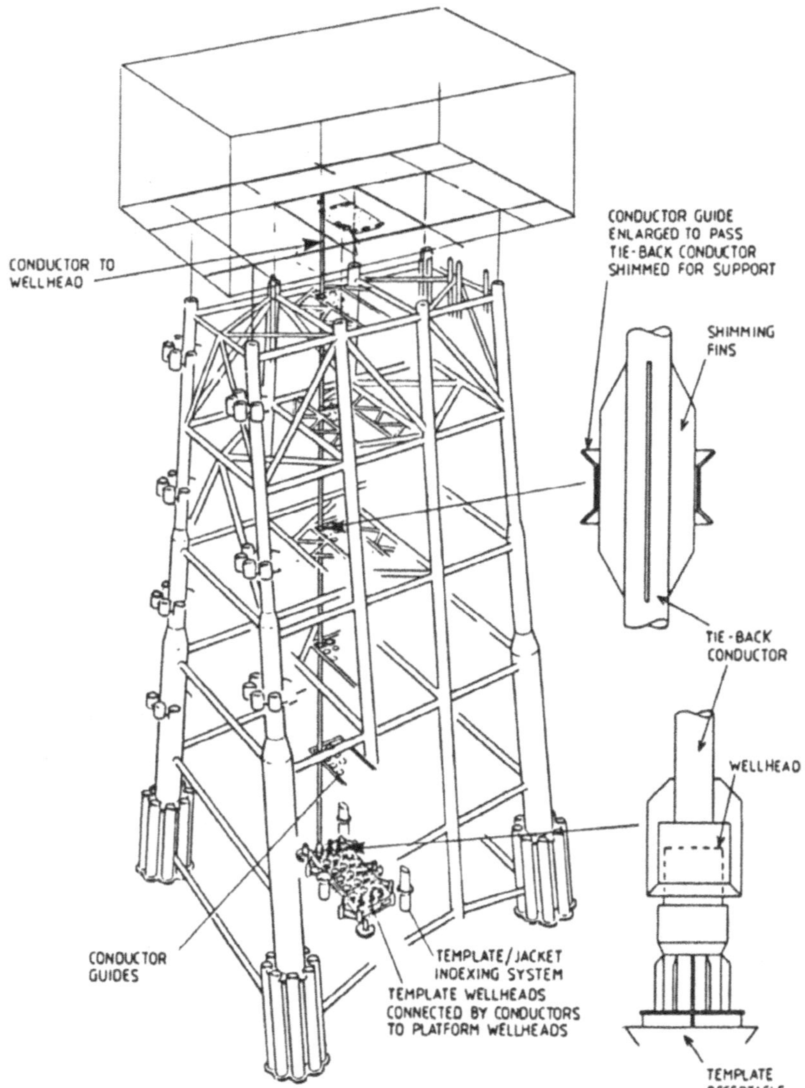

Fig. 4 Typical tie-back of the pre-drilled wells.

tubulars should be provided for protection purposes. Dependent upon the protection philosophy, a roof to protect the equipment against dropped objects may be provided. If a roof is not considered necessary, then a grating should be provided on the low-level base structure to prevent damage to the pipework.

It is a stand-alone facility which is controlled remotely, either from onshore or from a conventional offshore structure sited at a separate location. Produced fluids can be piped either directly to a shore terminal or to a production facility, usually located in shallower water, for processing before onward transportation. Other options also exist, such as production via a floating facility moored above the template or production into a storage tanker which can return to shore once its tanks are full.

TEMPLATE DESIGN

There are several considerations which affect the design of all templates, whether they are intended only for drilling or for both drilling and production operations. These requirements fall into the following areas:

- functional;
- structural;
- design;
- fabrication.

Functional requirements

The requirements of drilling templates can be summarized as follows:

- to space and arrange the wellheads so that there is sufficient distance between each wellhead to allow the blow-out preventer (BOP) to be fitted and for the wellheads to align with the tower conductors;
- to support the outer well casing before it is grouted in its oversized hole;
- to provide positive support to the wellheads at a suitable height above the sea bed, so that drilling cuttings and/or grout do not cover them;
- to provide facilities for guiding and docking of the production facility (whether it is a platform or a floating structure).

The production template should also provide the above (except for the docking facilities) and, in addition:

- means for installing production equipment;
- support for the production equipment;
- provision for tying-in of flowlines and service lines;
- a protective structure to prevent damage to the production equipment from externally applied loads.

The equipment on such a template would typically include production piping from the Christmas trees, tree controls and well-killing systems, ROV and diver aids, valves, manifolds and flowline bundles, and connection equipment systems.

Depending on the water depth and maintenance philosophy, a template system may be designed to be diverless or diver-assisted. In the case of diverless systems greater emphasis is placed on ROV access corridors for inspection.

Structural requirements

The aim of any template design is that it should be as small and compact as possible whilst still fulfilling the functional requirements. The main considerations which affect the template are outlined below.

Well layout

The configuration of a drilling template is dependent on how many of the total number of wells are to be pre-drilled. If the conductor arrangement in the platform structure consists of an 8 x 3 array but only 12 wells are to be pre-drilled, then the template could consist of a 6 × 2 array or a 4 × 3 array. Alternatively, a full 8 × 3 template could be installed which would allow greater freedom for the drilling operations. If the template does not house the full conductor array, then the location of its piles and supporting mudmats must be arranged to ensure no interference with any drilling activities undertaken after the platform has been installed.

Wellhead receptacles

Receptacles are required on pre-drilled wells to centralize the drill string and provide support for the wellhead housing and first casing before grouting.

Receptacles are typically tubes of 42 in. outside diameter from 1 to 3.5 m long with a cone at the top to assist drill engagement and a cone at the base to assist drill disengagement. There are two methods of suspending the wellhead housing within the receptacle, which are called the fixed and pinned systems. For the pinned or free-gimballing system the wellhead is supported at the top of the receptacle and is allowed to centralize within the receptacle itself. This system is adequate provided that the wellhead housing is not forced from the vertical by the drilling operation. As an alternative, the fixed or constrained system can be used, in which the wellhead is fitted within the receptacle to a close fit tolerance. In addition to providing vertical support, it also provides a moment restraint at the top and lower parts of the receptacle. A typical example of the arrangement for a free-gimballing wellhead is shown in Fig. 5.

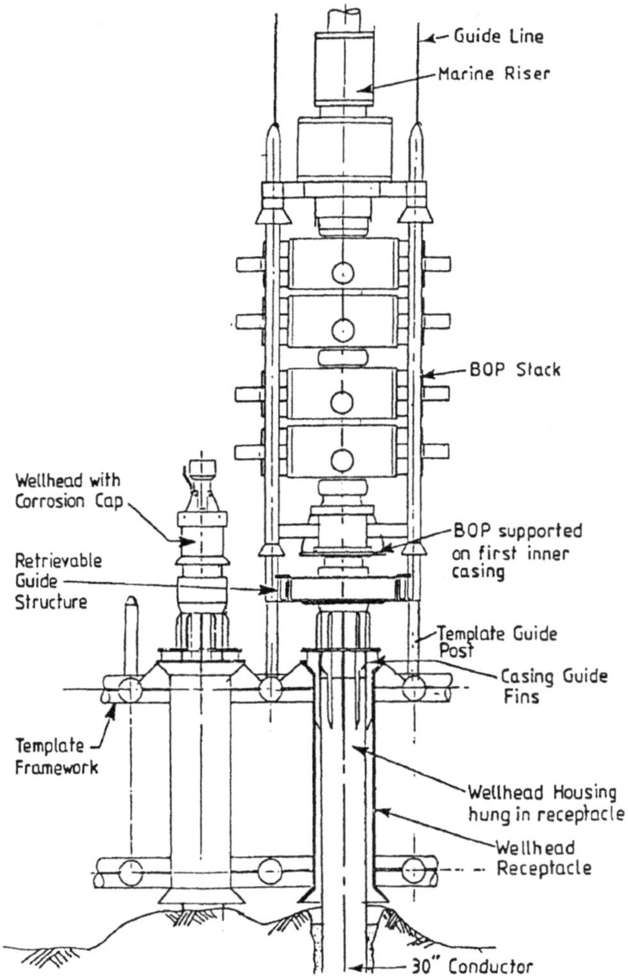

Fig. 5 Equipment arrangement at wellslot.

Framing options

The two main options available for framing of a template are the direct and indirect systems, both of which can utilize either plate girders or rolled sections. The direct system involves using the wellhead receptacles as primary structural elements of the template framing, whereas for the indirect system the main structural members bypass the receptacles, which have secondary members connecting them to the framework.

The direct system of template design results in a more compact structure with fewer members and less fabrication than a comparable indirect system template. However, the potential delays in obtaining the machined receptacles tend to negate these advantages, because of their effect on the fabrication schedule. Figure 6 shows some of the alternative framing options available for a drilling template.

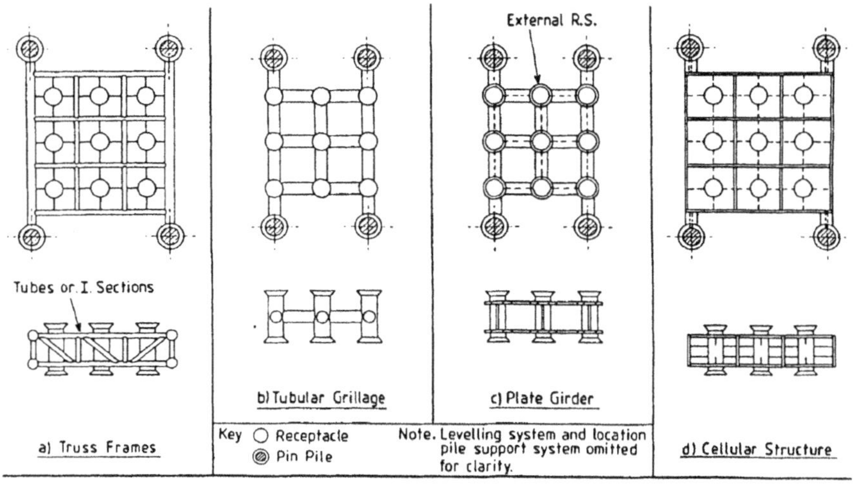

Fig. 6 Alternative framing options.

Jacket mating

If it is necessary to place a jacket over the template, then the use of locating or docking piles attached to the template will need to be considered. The location and number of such piles depends partly on whether the platform is to be mated vertically or horizontally with the template. Vertical docking involves upending the jacket with a high sea bed clearance and then manoeuvring it vertically over the template where location is achieved by the use of a mating pin and sleeve system. For the horizontal system the structure is upended away from the template system before being brought in horizontally. Location is achieved through a system of bumpers in addition to the pins and sleeves.

Whichever of these methods is selected, two factors are important: that the docking piles should generally be located outside the envelope formed by the remainder of the template and that they should be separated from it structurally before the installation of the platform structure. Both requirements are necessary to avoid possible damage to the template during docking manoeuvres. They are achieved by driving the docking piles through the sleeves at the extremities of the template and then cutting the connections between the sleeves and the remainder of the structure.

The minimum number of piles required to locate the jacket is two, but in the past a third pile has been justified on the grounds of redundancy, should one pile be damaged during the operation. They must resist jacket docking forces when the sleeves on the jacket are lowered on to the docking piles, and have adequate stick-up above mudline to ensure that docking occurs above the template. They should be located at maximum centres for optimum rotational control of the jacket, close to support piles, outside the template frame and with consideration to the jacket bracing.

Access requirements

If a template is to be operated without the use of divers, sufficient access must be provided to allow the ROV to manoeuvre in and around the structure. This can significantly increase the overall size of the structure, with the inherent effects on weight and cost.

Maintenance requirements

Auxiliary equipment mounted on production templates can include manifolds, flowline and umbilical tie-in equipment, control and distribution equipment, and pigging facilities. Removal and maintenance of the equipment is an important factor in determining equipment layout and structure size. Usually it is required that all equipment be vertically retrievable. Consequently, it is not possible to "stack" the equipment, and the structure plan size increases further.

Trawling protection

Trawling activities and water depth are linked factors which determine whether a production template will need to be protected from fishing activities. In very deep waters, where there is no fishing activity, it is not currently considered necessary to provide sloping sides on a structure to guide trawl wires over it. However, in shallow water, where fishing activities are common, the addition of sloping sides is a requirement which significantly increases the structure size. If the protection structure is intended to be snag-free and guide the trawl boards upwards over the structure then the sides will slope inwards ("A" frame). This configuration will also reduce the spans required for any roofing structures. However, if the protection philosophy is to catch the trawl wire and break the cables before it can approach the critical equipment, then a structure with the sides sloping outwards ("V" frame) can be provided and the provision of a roof will not be so critical.

If it is possible to define the area around the template as a fishing-free zone, then a protection structure with vertical sides should be sufficient. The protection structure should essentially consist of a ring of tubular members just above the height of the highest equipment, with sufficient capacity to withstand impact from the design accidental loads caused by trawling activities.

Installation vessels

If a heavy-lift vessel is to be used for the installation it is unlikely that it would impose any restrictions on the template size. However, limits may be imposed on the template size and weight if it were to be installed by a drilling vessel or jack-up due to clearance requirements or lifting capacities. If such a limit proved excessively restrictive, then the choice of vessel should probably be reconsidered.

Support requirements

Mudmats are required for temporary support and levelling on the sea bed until the support piles are installed and connected to the template. A levelling system is employed to ensure that misalignment of the well slots due to an uneven or sloping sea bed can be compensated for. The usual choice is between the redundant system with four jacking points and the statically determinate system with three jacking points. The mudmat locations should be chosen to ensure good control over the template level and minimum interference with the support piles and wells. As the jacks will almost certainly be a long-delivery item, a location-and-fixing system which will allow them to be added on at the end of the fabrication period is advantageous.

Support piles are generally considered essential for steel-framed templates, to provide verticality for the wellhead receptacles through the drilling phase. In addition, they should ensure that support points for the equipment do not undergo large differential changes in level which would make the interconnection of such equipment extremely hazardous. The number and location of the piles should be chosen to suit the overall shape and size of the template, bearing in mind the requirement for direct load paths and good template stiffness.

Design loads

During each phase of its life the template must be designed to withstand various combinations of permanent, live, environmental and accidental loads. The list in Table I, albeit brief, provides an indication of the critical load combinations for each major phase.

Table I

Critical load combinations for each major phase

Phase	*Description*	*Design*
Fabrication	Construction phase: various temporary support conditions	Conditions need to be checked after receipt of fabrication proposals. Unlikely to be critical to design
Loadout	Transfer from fabrication yard to transportation vessel	Skidding or moving on roller base: check for uneven supports and horizontal winching or jacking forces. Crane lift: check structure if temporary lift points to be used
Transportation	Transport to installation site	Design structure and sea-fastenings for accelerations due to expected seastates
Lifting and lowering	Lifting from installation vessel and lowering to sea bed (includes transfer from transporation vessel)	Lifting: design for skew effects, dynamic factors and include weight contingencies. Lowering: check for wave slam, consider added-mass effects
Emplacement	Landing on sea bed	Design structure for impact and uneven loading on mudmats
Levelling	Providing a level base for drilling and equipment	For three-point system check movement caused by overjacking. For four-point system check racking effects
Pile installation	Driving or drilling piles followed by connection to structure	Design for pile stabbing and overturning (both support and docking)

Drilling	Running and grouting conductor. Running and landing BOP	Check structure for support to conductor before grouting for each well. Check for percentage of BOP loading affecting structure. Note that support for structure changes after each well is grouted
Docking forces	Installation of jacket over template	Horizontal and vertical loads applied to indexing pile. Pile sleeve should have been previously disconnected from structure
Tie-ins	Pulling in the flowlines to the template	Design for winching forces
Tie-back	Connection of conductors to pre-installed wellhead	Check local design for alignment forces and moments. Tolerances dependent upon tie-back system chosen
Operating	Template producing gas or oil	Check structure with all equipment in place for wave and current loading, earthquake and fatigue Check for effect of thermal expansion of wellhead Check structure for direct snagging by anchor or trawl board or for flowline snagging Check for effect of dropped objects

Fabrication considerations

In order to achieve a cost-effective solution for any template structure, careful consideration should be given to fabrication requirements at the conceptual stage of design.

It is essential that the fabrication should be carried out as quickly as possible, so that the template can be installed at the earliest possible time during the weather window. This requires that the template should be simple to fabricate and that the materials and the vendor-supplied items should be available within a short period. By making the design simple to fabricate, it should be possible to subcontract the template fabrication to a wider range of contractors other than just the principal offshore fabricators. By including the smaller contractors, fabrication costs can be lower than for traditional offshore structures.

Section sizes for template base frames will, to some extent, be dictated by the overall height of the proprietary wellhead equipment and the need to provide support to these. This will generally result in the support members having low stress levels, because their overall dimensions will be governed by local stability criteria. Therefore, the choice of low-yield material would appear to be the optimum choice. In addition, by specifying a minimum grade with which most fabricators are familiar, a wider choice of tenderers should be available. However, for protection structures surrounding production templates, longer spanning members may be necessary to accommodate access requirements. For these members high-yield steel may be essential to allow for design against snagging loads. In this instance, if high-grade steel is readily available without a cost penalty, then its use throughout the structure should be considered.

It is sometimes a requirement for the wellhead equipment vendor to supply the receptacle. This may be necessary to ensure that the vendor is satisfied with the fit-up tolerances of the housing within the receptacle before actual fabrication into the template. In addition, it would be beneficial for the vendor to attach any stub section required for connection into the template before machining the internal bore of the receptacle. In this way any weld distortion will be removed during the machining process. Subsequent connection into the template structure would therefore not affect the internal tolerances of the receptacle.

TEMPLATE INSTALLATION

Installation may be regarded as the activities occurring from fabrication until the structure has been piled in place on the sea bed. For large subsea templates the installation condition may, indeed, impose the most severe loads experienced by the structure. The major factors to be considered for design of the structure and installation aids are discussed in the following sections. A summary of the various options for each phase of the installation is given in Table II.

Loadout

The choice of loadout of the structure will depend upon the fabrication yard facilities, and will normally be performed by one of the following methods:

Table II
Options available for template installation

Activity	*Options*
Loadout	Lift, roll or skid
Transportation	Barge, crane vessel, drillship or semi-submersible
Lift and lower	Conventional four-point lift, spreader beam or frame, auto or manual release of rigging. Hard or soft slings, buoyancy assistance
Positioning	Tugs, tuggers with sea-bed anchors, tuggers from installation vessel, thrusters
Level measurement	Acoustics, spirit level, bullseye
Level adjustment	Jacking from mudmats or piles
Support-pile installation	Driven or drilled. Pre-installation of piles in template or lowered with hammer. Pile followers or underwater hammer. Pile cutting
Support-pile attachment	Swaged, grouted or mechanical
Docking-pile installation	Hammer requirement and mode. Piles may be lowered alone, with hammer or in template
Removal of docking-pile guide	Diver or ROV. Quick release or cut connection

- skidding;
- rolling;
- lifting.

It should be possible to move the structure from its fabrication site to the transportation vessel by any of these methods without excessive strengthening. However, the following points should be considered when determining which loadout method is most appropriate for a given structure:

- position of structure in yard;
- size of barge or vessel to be used;
- required position of structure on the barge;
- access from fabrication position to barge location;
- water depth adjacent to the loadout quay and access for transportation vessel;
- mooring arrangements;
- structure loads;
- limiting environmental criteria for the selected loadout method.

Transportation

Transportation concerns the movement of the template structure from the dockside to the installation site on the transportation vessel, where it may be transferred to the installation vessel. A supply boat or an unpowered flat barge may be used, or a crane vessel with sufficient deck space can be chosen.

The final choice of vessel will be dependent upon the size and weight of the template plus any associated items required for the installation, such as piles, rigging equipment and lifting aids. These items should be pre-installed and sea-fastened on to the barge so that the final action is to install and sea-fasten the template.

Template installation is a weather-sensitive operation, as bad weather prevents the lowering of the template through the wave zone or inhibits pile installation. Vessels which exhibit small motions in bad weather are preferred, but these tend to be costly to hire and a balance between vessel motion and hire cost is often required.

The types of vessels which could be used for the transportation and installation are discussed below.

Semi-submersible crane vessels (SSCV)

This type of vessel is ideal for the handling and lifting of templates. They offer large deck areas suitable for both the template and the miscellaneous equipment needed for installation. In addition, because of their good motion-response, they form a stable platform that is preferable during installation operations.

The principal vessels in this category are all capable of lifting more than 2000 short tons (1815 tonnes) with their main block and more than 500 short tons (454 tonnes) with their auxiliary block, and are capable of picking up a typical template from the fabrication yard, if it has a deepwater quay, or from a transportation barge in sheltered waters.

When the SSCV is positioned at the site the template can be lifted from the deck of the vessel by slings connected to the main block, moved over the side and then lowered, preferably using an underwater block.

Semi-submersible drilling vessel (SSDV)

A drilling rig can be used to transport and install a template by attaching it under the deck, or lower-hull bracing, provided sufficient clearance is available between the legs of the rig.

Most drilling rigs have a maximum static load capacity of about 500 short tons (454 tonnes), whilst the draw-works may have a lower capacity. It is believed that a load of about 300 tonnes is the practical maximum which the derrick can lift. However, this will vary from rig to rig, depending on the make of draw-works.

During transportation the possibility of damage to the template by wave slam could lead either to a restriction on the operable seastate, or to the need for further stiffening of the structure. The rig transports the template to site and, on arrival, the sea-fastenings are cut and the template is lowered to the sea bed. To lower the template additional lengths of riser pipe are run down until the template is located on the sea bed. The rig is compensated for heave motions, so lowering operations should be reasonably straightforward.

Heavy-lift vessel (HLV)

HLVs are essentially built for lifting and installing offshore structures. They can be used for transportation, but ideally the template would be taken to site on a flat-topped barge.

A typical ship-shaped HLV has a 2000 short tons (1815 tonnes) main block and a 400 short tons (363 tonnes) auxiliary block. It can be used for lifting templates from a transport barge, either in sheltered waters or on site, or if there is sufficient deep water, at the fabrication yard.

Many of the advantages of the semi-submersible crane vessels apply to the ship-shaped HLVs, and they generally command a lower day-rate. However, their heave and roll characteristics are not as good as those of the semi-submersible vessels, which can lead to longer and more-complex installation operations.

Jack-up rig

A jack-up rig can be used for the installation of a template in a similar manner to that of a semi-submersible drilling rig. Although limited to the water depth in which they can operate, they have the advantage of a wider spacing between the legs and a more stable work platform over semi-submersible drilling vessels.

Flat-bottomed crane barges

Although these barges have the capabilities of lifting and handling templates, poor motion characteristics result in a low operability and also increase the difficulties of handling operations. However, given the right weather conditions, a flat-bottomed crane barge could be a very economical solution.

Lifting and lowering

If the template is to be lifted offshore from the transportation vessel by a heave-crane vessel, then typically one of the following arrangements will be used.

Conventional four-sling lifting arrangement

The configuration for this lift would consist of a pin, or shackle, through a padeye, with four slings attached, leading back to the main lift cable. The inclined slings impart lateral thrust into the template structure as well as racking loads.

Spreader-frame assisted lift

This system makes use of horizontal framework which normally has four padeyes. The frame removes the horizontal thrust components of the sling loads from the template, thus making the lift forces less onerous upon the template structure.

Spreader-beam assisted lift

In a similar manner to the frame-assisted lift, a spreading beam can be utilized. It will only remove the horizontal thrust in the axial direction of the beam, and can be used to advantage if the template has a weakness about one axis only.

The main criterion for the choice of the lifting system is that it should enable the template to be lifted, lowered and placed without member overstress or excessive deflection. However, unless the use of the conventional four-sling arrangement results in excessive requirements in terms of weight of steelwork for the design of the template, then use of a beam or frame should be avoided. Consideration should be given to pre-rigging the slings to avoid the operational difficulties that can be encountered during the connection of the slings on-site due to significant motions.

The lifting system should place the lift block as near as possible over the centre of gravity of the structure to ensure a level lift, since any eccentricity of the lift block relative to the template centre of gravity at the start of the lift may result in a lift-off which is not level. Apart from creating uneven loading in the lifting equipment, an out-of-level template at touchdown on the sea bed could result.

The forces on the template structure after lowering through the splash zone arise from the submerged weight of the structure plus any dynamic effects, and the acceleration of the structure relative to the water, which can develop large added-mass loadings.

The criterion of lowering the template underwater is to ensure that the dynamic response of the structure to wave-induced crane motions does not overload the slings. If the crane-hook frequency coincides or approaches the natural frequency of the lowering system, then the structure dynamic amplitude may increase considerably with respect to the crane-hook amplitude. The effect can significantly increase the forces experienced by the template during the lift.

Positioning

It is essential that the correct location and level be provided for the template during installation.

Accurate positioning of the template on the sea bed in a predetermined orientation requires a system to monitor the position of the template as it is lowered, and a control system to adjust the template position and orientation. This is especially important as it nears the sea bed.

To monitor the template position, an acoustic system may typically be used. This involves establishing a pattern of transponders on the sea bed in the installation area, then locating a survey vessel, carrying navigation equipment, inside the pattern. The vessel position is then established relative to fixed datum points and then to the transponder. This results in calibration of transponder location on the sea bed to around 1.0 m. Using two more transponders on the template allows the template to "home" on to the sea-bed transponders via the survey vessel, and this enables the template position and orientation to be determined. The survey vessel can be eliminated by locating the survey equipment on the installation vessel.

Final touchdown of the template on the sea bed may be assisted by visual monitoring using a ROV/RCV or a diver-held camera. To adjust template position during lowering there are numerous options available including:

- control lines from the lift vessel;

- control lines from tugs;
- tugger lines via sea-bed anchors.

Control lines from the lift vessel could result in such acute angles at the template that at the sea bed, where control is most needed, the vertical component of line tension would exceed the horizontal component and could seriously affect the template level. This method has, however, been used successfully in previous operations, but its choice is dependent upon environmental conditions and the particular template profile.

The use of control lines from tugs may require such a large catenary for the installation depth that line response would be greatly impaired and sensitive to weather and tug communications. Their use adds to the expense of the offshore operation and is difficult to control.

To utilize a sea-bed anchor system requires pre-installation of anchor piles and control lines. Usually four anchors are pre-installed on the sea bed, with the lines buoyed-off ready for use. This method requires lines to be heave-compensated on the installation vessel, but has the advantage that tugger efficiency increases as the template nears the sea bed, where control is most needed.

Levelling

Typically, templates are levelled by hydraulic jacks acting against mudmats or the support piles.

Where piles are used, pile slips can be located on the pile sleeve, which allows the template to be raised (but not lowered) by a jacking tool. The tool is run down on a drill string and fastens to the top of the pile sleeve. When the hydraulic ram is actuated, it pushes down on the pile top, and the template consequently rises.

Where levelling is performed by mudmats, the template is lowered to the sea bed with the hydraulic jack extended to half the maximum travel, to allow initial levelling by pressure release alone. A hydraulic line from each jack can run to a common manifold on the template which is, in turn, connected to the surface via a single umbilical. The umbilical can be either lowered with the template or attached subsea by a ROV. The ROV can then allocate hydraulic control to individual jacks as required.

By using a three-jack arrangement the only effect of overjacking on one mudmat would be to cause a rigid-body movement about the line of the other two jacks, and this will not introduce any significant additional stress in the template.

If a four-jack arrangement is adopted, then overjacking on a single mudmat can lead to the situation where virtually the whole load is supported on two diagonally opposite mudmats, severely racking the whole structure and imposing high lateral loads on the jacks.

Although a three-jack system is considered more desirable, if the three jacks cannot be positioned such that an even load distribution is obtained or the template levelled to sufficient accuracy, then the use of four jacks and mudmats should be favoured.

Template level can be monitored from the installation vessel via an electrical umbilical to liquid-level sensors (U-tubes) located on the template. Failure of the umbilical connection can be compensated for by direct read-out by ROV of the U-tube, and even total failure of these can be allowed by observation of the less accurate "Bullseye" level indicators positioned on the top of the template.

Piling

Support piles are required to locate the template positively so that it can resist, with minimal vertical or horizontal displacement, the effects of scour and the loadings resulting from docking, drilling or operating activities. Dependent upon the soil conditions, a choice can generally be made with regard to drilling or driving the piles. The requirement for a drilling rig for piling could be combined with its use as an installation vessel to produce a more economic solution. However, in soft or granular soils drilling "wash-out" or "cave-in" of the drilled hole can be a significant problem that precludes drilled piles. The consequences of either of these problems occurring would be a greatly extended installation operation with the requirement for additional grouting to prepare the hole for the pile.

"Wash-out" or "cave-in" could also seriously impair the template stability if the mudmats are undermined. The decision to undertake either the driven or drilled operation will be based on a study of the economics of both, including a consideration of this problem.

Having installed the piles, the next requirement is to connect them to the template. There are three principal methods of making the connection between the pile and pile sleeve, and these are noted below together with their advantages and disadvantages. The choice of system will, of course, depend on the exact requirements of each template, so a careful economic assessment of the options will be required.

Grouted connection

In this method grout is pumped from a surface vessel through an umbilical into the annulus between the pile and pile sleeve. Shear connectors welded on to the pile wall and the inner sleeve transfer the load into the grout. This is a proven system that has been applied to a large number of pile-to-sleeve connections. The fact that no machined parts are required helps to avoid both fabrication problems and operational difficulties. Special equipment in the form of packers to contain the grout inflation lines for the packers, grout lines from a central grout umbilical and grout-density monitors may need to be provided.

Swaged connection

Once the pile has been driven to its target penetration, a short length of the pile wall is radially deformed into grooves machined in the pile-sleeve wall. The deformation is achieved by cold-forming of the steel by the application of very high hydraulic pressure inside the piles. The special tools required for the swaging operation are currently available for 16, 24 and 30–42 in. piles. It has been used on a number of templates, although it has still to be considered a new technology. The annulus between the pile wall and machined sleeve is of the order of a few millimetres, so that pile-head damage during pile lifting or driving must be avoided.

Mechanical connectors

In this system wedges or slips are welded to the outside of the pile wall and to the inside of the pile sleeve. The pile is driven to a precise penetration, wedges then being screwed along the pile axis to give contact. This is a direct mechanical-load transfer system.

4

Flow bundles and developments in pipeline installation

M. H. Chancellor and D. A. Phillips,
J. P. Kenny & Partners, Aberdeen

INTRODUCTION

The recent reduction in oil prices throughout the world has led to an increase in the number of offshore oil and gas fields attracting marginal status. The reduction has led many operators to reconsider the conventional offshore development scenario involving above-water production facilities and associated export pipelines. This revaluation has led to a significant increase in the number of subsea satellite developments, together with the accompanying transport and service lines tied back to existing production facilities.

This approach often allows for maximum utilization of existing investments by the addition of further throughput achieved by a new, relatively low-cost satellite development.

Currently, the lack of available subsea process facilities demands that wellhead flowlines be manifolded and extended back to the "parent" facility. Typically, this may result in a number of flowlines between the satellite development and existing facility to provide services such as oil and gas production, well test, gas lift, and chemical and water injection.

Some examples of development schemes of this type currently in production or being constructed are as follows:

Advances in Underwater Technology, Ocean Science and Offshore Engineering, Volume 11 – Underwater Construction: Development and Potential

- Texaco: Highlander Field tied back to Tartan (North Sea, UK Sector);
- Occidental: Scapa Field tied back to Claymore "A" (North Sea, UK Sector);
- Statoil: Tommeliten Field tied back to Edda (North Sea, Norwegian Sector);
- Mobil: Ness and BBSWI tied back to Beryl "B" (North Sea, UK Sector).

In contrast with conventional above-water developments, this new generation of offshore oil and gas fields may involve the installation of a significant number of small-diameter, relatively short-length pipelines.

This feature requires that the use of the conventional high-cost third-generation lay-barge, developed for deepwater installation of larger-diameter pipelines be re-evaluated. Consequently, detailed consideration should also be given to alternative methods of pipeline installation. Moreover, as the provision of pipelines forms a large proportion of the capital cost of a satellite development, it is necessary for the pipeline designer to evaluate carefully all available installation methods early in the design process, in order to identify the most cost-effective solution.

This chapter presents a brief review of the requirements for pipelines for this application of satellite developments, and the suitability of various pipeline design and installation methods.

GENERAL DESIGN CONSIDERATIONS

Design requirements for traditional offshore pipelines are well documented and are not addressed here. However, where multiple small-diameter pipelines are envisaged for satellite developments, special consideration of the following areas may be required during pipeline design:

- material selection;
- thermal insulation;
- installation – pipeline corridor requirements;
  – tie-ins;
- deepwater installation and tie-in.

Internal corrosion may result from high hydrogen sulphide or carbon dioxide levels in the produced reservoir fluid. However, state-of-the-art technology relating to subsea process equipment is not yet sufficiently developed to influence internal corrosion in satellite field export pipelines. Consequently, selection of more-exotic materials, such as duplex stainless steel or carbon steel clad with stainless steel, may prove to be more attractive than the use of an anticorrosion chemical-injection system. Selection of these higher alloys necessitates detailed consideration of the installation method during design, with regard to their weldability, ductility, material capability and external corrosion resistance in the seawater environment.

For example, it is recognized that difficulties exist in the welding of duplex stainless steel line pipe, resulting in joint welding times typically in the range of 40–60 min. As this will result in a significantly reduced lay rate, and consequently increased lay costs, there exists a high incentive to fabricate the pipelines onshore in a controlled lower-cost environment. However, subsequent installation of these prefabricated pipelines must be addressed with regard to their material properties as discussed below.

Similarly, the lack of subsea processing equipment can result in a requirement to insulate pipelines to inhibit hydrate or wax formation at critical flow conditions. Thermal insulation is typically achieved by the addition of low-density coatings, the requirements for which are significantly affected by the installation method selected. For example, the encasement of pipelines within a carrier pipe to form a pipeline bundle will result in a lower insulation requirement, due to the overall reduction in heat loss. This is attributable to the insulating effect of the medium contained in the annulus, between the bundle and the inside of the carrier pipe.

Furthermore, the introduction of gels rather than inhibited water into the annulus will inhibit convection currents and may provide sufficient thermal insulation without additional coatings.

Multiple pipelines from a satellite development can result in a wide route corridor which may not be acceptable with respect to vessel-anchoring restrictions or fishing-authority requirements. Furthermore, the tie-in of pipelines laid over a wide corridor may result in unacceptable congestion, particularly at the base of an existing platform. Consequently, the grouping of pipelines into bundles or "piggy-backing" smaller-diameter pipelines together may lead to a more acceptable corridor width and more control of the overall tie-in requirements.

Detailed consideration of installation and tie-in of pipelines in deep water, outside the currently acceptable range of diver intervention, is not addressed in this chapter.

However, some of the features of the remote tie-ins considered for these developments may have application for multiple-pipeline tie-ins in water depths suitable for diver intervention. Specifically, the guiding of the pipeline or bundle ends by the use of pre-installed mechanical aids will assist in the control of end tolerances, and may lead to a reduction in the overall tie-in schedule.

COMPARISON OF INSTALLATION METHODS

Several discrete installation methods for offshore pipelines exist. Those suitable for the installation of small-diameter pipelines are summarized below:

- conventional S-lay barge;
- reel barge;
- new-generation lay vessels;
- towed bundles.

Conventional S-lay barges

Conventional S-lay barges have successfully installed many kilometres of offshore pipelines, and can without doubt adequately install sets of small-diameter pipelines. However, these vessels have been primarily developed for the economical installation of relatively long pipelines with diameters which in general are significantly larger than the typical 2–10 in.* range associated with many satellite developments. Consequently, alternative installation methods may prove to be more economically viable where the total project pipeline requirement is measured in tens rather than hundreds of kilometres, especially when this total may consist of several short lengths of different diameters.

Reel barges

As an alternative to conventional S-lay vessels a number of reel vessels have been built, the most notable of which is the *Apache*, operated by Santa Fe. This is claimed to be capable of laying pipe up to 16 in. in diameter, but more significantly it is ideally suited to smaller-diameter lines, which are strained to a lower level when wrapped around a large-diameter main reel. Furthermore, this

*1 in. ≈ 2.54 cm.

vessel has the advantage of dynamic positioning, which allows for rapid location at the pipelay initiation point and for subsequent travel along the pipeline route, unhindered by the time-consuming requirement to set and regularly reset an anchor spread. It also has the added advantage that potential problems of anchor deployment in congested areas are eliminated.

New-generation lay vessels

Historic development of lay vessels up to the current third-generation equipment, such as Bar 420 (ex Semac 1), Castoro Sei and LB200, has been aimed at the efficient laying of relatively long lengths of larger-diameter pipelines.

Today, a considerable incentive exists in the development of a new generation of lay vessels which will be more cost-effective in laying smaller-diameter pipelines in short lengths. Coupled with this pipelaying capability, the vessel specification should include a trenching facility, saturation diving spread for performing tie-in and possibly a mid-range lift capacity for the installation of small subsea structures.

The most notable example of this development to date is the Allseas *Lorelay* which has the following capabilities:

- dynamic positioning;
- suitable for installation of 4–24 in. diameter pipe in water depths up to 360 m;
- substantial double-joint pipe-carrying capacity via cargo hold (e.g. 12 km of 20 in. pipe or 40 km of 6 in. pipe);
- 16 knot transit speed;
- six single-joint or three double-joint automatic welding stations;
- revolving 120 tonne crane at 30 m radius;
- provision of support services, including trenching, ROV inspection, pipeline tie-in and repair.

Conversion of existing diving support vessels to perform pipelay is gaining increased interest since it was used recently in Morecambe Bay with the laying of a 2 in. pipeline by the BAR *Protector*. Recent studies have indicated that S-lay, J-lay and reel packages are suitable for installation on to a number of larger DSVs, and would provide a pipelaying capability of up to 12 in. diameter in the medium water depths found in the middle and southern regions of the North Sea.

Both S-lay and J-lay, from a vessel of this type, will benefit from the development of an acceptable rapid-welding facility, which will allow reasonable lay rates to be achieved with only one welding station in operation.

The addition to small vessels of portable reeling packages with a horizontal reel axis severely limits the overall size of the reel, and the consequent pipeline capabilities, due to the unacceptably high overturning moments generated.

Clearly, the use of a vertical-axis reel (horizontal reel) will provide increased pipe diameter and pipe-carrying capabilities.

Towed bundles

As an alternative to laying pipelines individually, they may be grouped together within a carrier pipe as a pipeline bundle and towed offshore. This method of installation has been utilized successfully on a number of developments throughout the North Sea and considered in detail for developments in the Gulf of Mexico and offshore Australia.

Examples of successfully installed bundles in the North Sea are:

- Occidental Claymore Field extension (water injection);
- Shell underwater manifold centre;
- Occidental Scapa.

Flowline bundles fall within two broad categories:

- strapped bundles;
- towed bundles (within a carrier pipe).

Strapped bundles typically consist of one or more small pipelines attached to a larger-diameter pipeline. These are installed by conventional S-lay vessels or by reeling, and have the advantage of significantly reducing offshore construction schedules, where appropriate. However, control of the individual pipelines becomes difficult and typically this type of bundle would still require trenching to provide protection and on-bottom stability for the individual pipelines in the event of strap failure.

Alternatively, several pipelines may be grouped together within a carrier pipe to form a flowline bundle, which would be fabricated onshore and towed to the installation site. On completion of onshore fabrication the bundle is subjected to full integrity testing, including a hydrotest. Once testing has been finalized and the bundle shown to be satisfactory, tow operations may commence.

Tow operations begin with launch of the bundle from the construction site. The bundle is then transported offshore using the selected method of tow. Tow operations are terminated once the bundle lies on the sea bed at the offshore site ready for tie-in. These various aspects of the tow are used to develop vessel requirements, which are discussed later in this section.

The various methods of tow are described below such that familiarization with their specific techniques may be gained.

Tow methods may be characterized by the level within the water that the bundle is transported, i.e. at the surface at mid-depth or on the sea bed. These are referred to, respectively, as surface tow, mid-depth tow and on-bottom tow.

Each method has its own buoyancy requirements, and these can be provided in several ways, as described below. By varying the degree and type of buoyancy provided, variations on the above tow methods may be achieved. This results in the development of the following six techniques, which are described below:

- bottom tow without added buoyancy;
- bottom tow with added buoyancy;
- off-bottom tow;
- mid-depth tow;
- near-surface tow;
- surface tow.

Bottom tow without buoyancy

This is the most basic tow technique, and may be considered as a development of the pull technique used to construct pipeline crossings over rivers and estuaries, and to install near-shore loading lines and sewage outfalls.

With this technique the line is towed to site in its final configuration without the aid of additional buoyancy, as illustrated in Fig. 1. Consequently, this method is only applicable to strapped bundles.

During tow-out the bundle is dragged across the sea bed, the strapping holding the bundle together being subjected to abrasion and possible damage from impact with sea-bed obtrusions such as rocks or even other pipelines. The resulting risk to the integrity of the bundle is considered unacceptable. This method of installation is therefore not considered suitable for bundle application, and is not addressed further.

Bottom tow with buoyancy

This variation of the bottom-tow technique is not illustrated in Fig. 1. The pipeline system is pulled along the sea bed, as with the basic

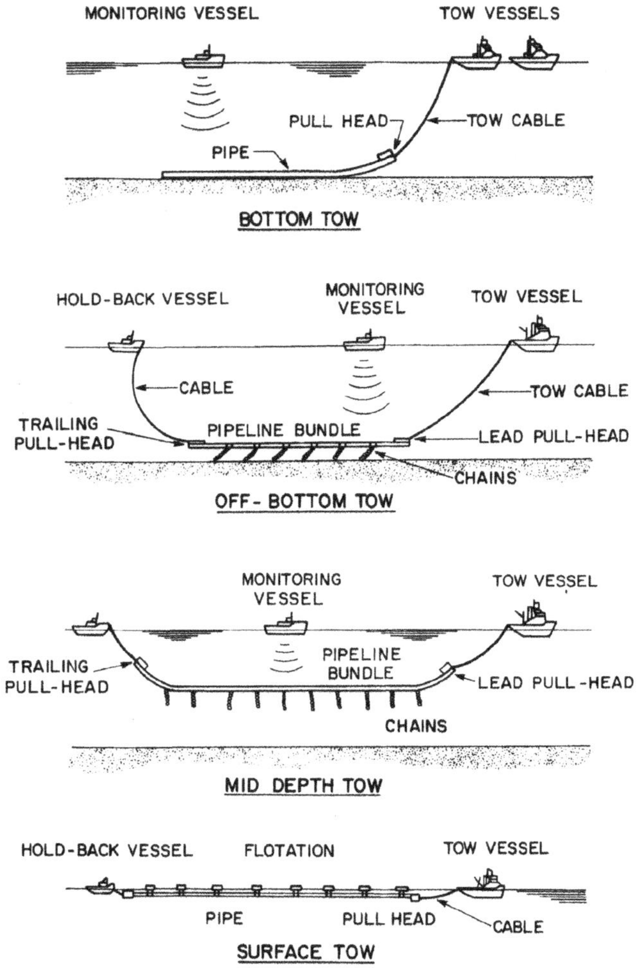

Fig. 1 Tow methods.

bottom-tow technique, but not in its final installation form.

With this method the system is provided with additional buoyancy in order to reduce the submerged weight. Although the bundle remains in contact with the sea bed, the force required to overcome friction is less than that required for a similar system without buoyancy. As a consequence this will allow either the use of a smaller tow vessel, or will enable longer pipeline systems to be towed. The additional buoyancy may be provided either in the form of auxiliary buoyancy tanks or by constructing the pipeline inside a carrier pipe. As the bundle is dragged across the sea bed, there is a tendency for a furrow to be cut along the direction of tow, especially in areas where surface soils are soft or loose. The majority of the bundle is therefore effectively towed along a shallow trench, and the likelihood of lateral drift from cross-current drag is generally reduced. The depth of the furrow will obviously vary in accordance with soil conditions and the submerged weight of the bundle, which will vary resistance to lateral drag. Should it become necessary (for reasons of vessel availability or bundle length) to reduce the submerged weight greatly, then it is likely that a trailing vessel will be required to control the bundle during tow. This requirement may be met by the survey vessel or a small tug.

A trailing towhead is provided on the bundle for manoeuvring into position at site, and this also provides a facility for attachment of a trailing tow vessel should it be required. By attaching a buoyed cable to the trailing towhead, such attachments may be readily achieved without the need for diver intervention.

On arrival at site, long-term sea-bed stability against extreme environmental conditions is maintained by removal of the buoyancy tanks or flooding of the carrier pipe. Such details are discussed in later sections.

Off-bottom tow

The off-bottom tow technique is a further derivation from the basic bottom tow. In this case the submerged weight is not just reduced, the system is designed to be positively buoyant. This buoyancy is then balanced by hanging trim, or drag chains, which are of a weight and length which suspends the bundle at a specified height above the sea bed. Again, the additional buoyancy may be provided by auxiliary buoyancy tanks or by constructing the bundle inside a carrier pipe. By providing an additional length of chain which rests on the sea floor during tow, a degree of vertical stability is introduced. This chain arrangement is self-compensating, such that in the event of the pipeline being lifted higher than its designed position, perhaps by excessive tow tension, the additional chain being lifted from the sea bed restores equilibrium. The chain length also compensates for variation in uplift due to changes in sea-water density and temperature.

As this method intentionally provides minimum frictional resistance against the sea bed, it has little inherent directional stability. A holdback vessel is therefore required for control of both the tow configuration of the bundle and forward movement. This method of tow is illustrated in Fig. 1. The general characteristics of tow by this method mean that the bundle is vulnerable to cross-currents. Such currents will induce drag, forcing the bundle sideways. This sideways movement is controlled by the leading and trailing vessels, which provide tension in the system to hold the bundle in a shape similar to a catenary. In accordance with the magnitude of the current velocity, the tension provided by the vessels is adjusted. In a stationary condition both vessels may maintain similar pull forces in opposite directions, which can be varied as necessary.

To provide forward momentum, the leading vessel increases its pull force with the trailing vessel simultaneously reducing its pull force. The bundle will then move forward, the velocity of tow and catenary configuration being controlled by adjusting the difference between the vessel pull forces.

Mid-depth tow

The mid-depth tow method is also illustrated in Fig. 1. It is a variation of the off-bottom tow technique, submerged weight and buoyancy being balanced in a similar manner. With this method, however, the pipeline is towed well clear of the sea bed.

In shallow water, or when the bundle is stationary, it is in essence an off-bottom tow system and control in cross-currents is provided in the same manner. During normal tow in relatively deep water the bundle is held at a depth of about 30 m. At this depth the effect of surface waves are minimal and will have no adverse effect. The required uplift is achieved by introducing greater tension between the lead and trailing tow vessels, together with higher tow velocities, both of which induce the system to lift off the sea floor.

The characteristics of the pipeline configuration during tow may be altered to suit the water depth and current variations by control of tow tensions and forward velocity.

The configuration taken up by the bundle and rigging during tow can be predicted by analysis using standard catenary theory. For this method of analysis to be applicable, however, the system must approximate to a cable. This can be checked by the use of the following formula which provides a "rule-of-thumb" guide based upon the stiffness (EI) of the bundle.

$$EI << T^3/W^2$$

where E is elastic modulus, I is moment of inertia, T is in-line tension and W is submerged weight per unit length. (*Note*: all units must be compatible.)

Should the stiffness of the system (EI) be greater than T^3/W^2 then stiffened catenary theory must be used.

By combining the effects of the submerged weight of the system with cross-current drag, the catenary analysis may be used to predict typical tow configurations and tension requirements to combat the likely environmental conditions. Such data are used to aid selection of vessel power requirements and to provide a guide to the tow master so that the required tow depth may be maintained. This is important in two directions. First, should the tension be too high, then the bundle will rise to the surface where it will be subjected to the effects of waves. This may lead to overstressing of the bundle and possible failure. Secondly, should the tension be too low, then the catenary will drop and the bundle may revert to off-bottom tow. Whilst this may not present a problem in relatively shallow water, in deep water or under high-current conditions the curvature of the bundle at the bottom of the catenary may be reduced to such a degree that overstressing may again occur.

Whilst it is expedient to monitor the bundle configuration carefully at all times, in general overstressing of the bundle does not present a serious problem, except under extreme conditions, as discussed above.

One further consideration which modifies the catenary analysis slightly is the drag on the bundle as it is towed through the water. The effect of the components of this drag is illustrated in Fig. 2.

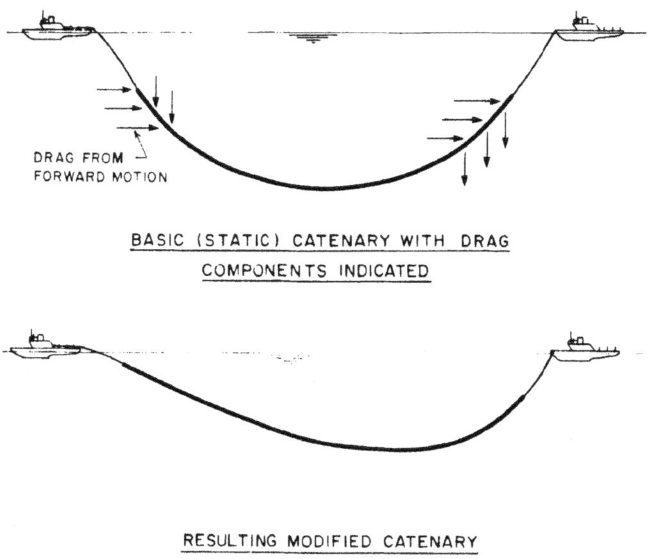

Fig. 2 Tow-out catenary configurations.

Near-surface tow

This method is illustrated in Fig. 1, and involves towing the pipeline system just below the zone of severe wave action. A typical depth is of the order of 15–20 m. This depth may vary, however, in accordance with predicted wave criteria along the route.

With this technique buoyancy is provided by spar buoys which are designed to suspend the bundle at the required depth. Typical spacing of the spar buoys is between 100 and 150 m, although this is subject to stress analysis related to the bundle/pipeline parameters and available wave data. Although other types of buoy may be used, the spar buoy has far less resistance to wave movement than a drum or pontoon, which will be subject to upthrust as the wave passes.

As the bundle is near the surface, it will be subjected to the maximum effects of current, including tide and wind components. As a consequence there will be a tendency for the tow configuration to be in the form of a horizontal catenary which may be predicted in a similar manner to that described for mid-depth tow. Again, the forward momentum will modify the catenary shape, which will take the form illustrated in Fig. 3. This figure illustrates how the heading of the lead tow vessel is adjusted to overcome the effect of the current pushing the bundle sideways such that the direction of the tow is maintained.

Surface tow

This technique is illustrated in Fig. 1. The pipeline system is designed with sufficient buoyancy to support the pipeline on the water surface. The required buoyancy may be achieved by constructing the pipeline inside a carrier pipe or by using buoyancy tanks. For this technique the buoyancy tanks must be closely attached to the bundle, unlike the spar buoys of the near-surface technique.

This system is influenced by surface currents to a slightly greater degree than the near-surface tow. More importantly, the surface tow is greatly affected by the waves themselves. Due to the buoyancy characteristics of the system, there is no inherent capability to compensate for wave action. This may lead to the bundle trying to follow the profile of the passing wave, which may result in overstressing.

A regular seastate is unlikely to occur with sufficient frequency that resonance will result. The variation in stress levels resulting from wave movement may, however, lead to fatigue failure. Under extreme wave conditions it is possible that resulting stresses may cause buckling of the bundle, leading to failure.

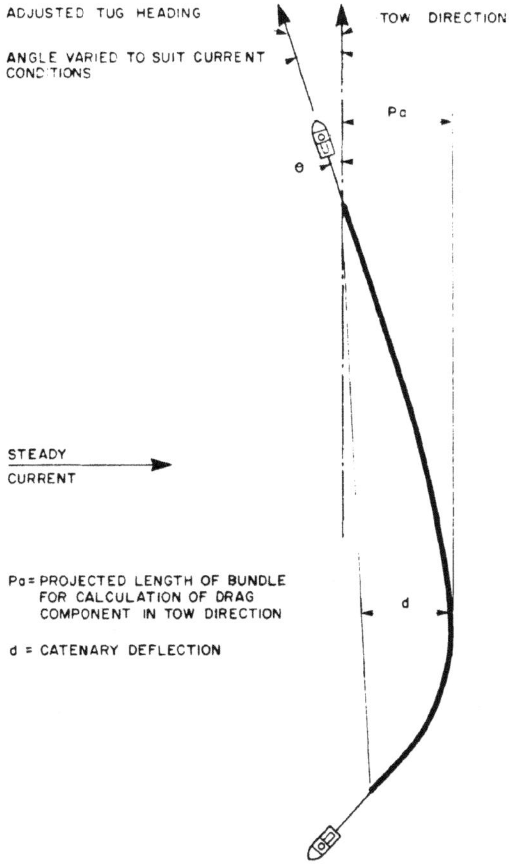

Fig. 3 Typical near-surface tow-out configuration.

A basis for tow-method selection

It is difficult to make specific recommendations for selecting a method of tow, as every installation has its own requirements. There are, however, some general considerations, as summarized below, which outline the merits and restrictions of each technique.

- Surface tow has inherent problems associated with wave action. As a consequence, this technique is normally considered suitable only for short lengths of pipe, or where wave action is minimal.
- Near-surface tow is generally suitable for any length of bundle. As there is some effect from wave action, however, difficulties may occur should this option be used in waters where summer storm waves may be excessive. Such areas may be found within the northern North Sea.

Another aspect of this method, which would also be applied to surface tow, is that complex procedures are required to install the bundle on the sea bed. This must be achieved either by controlled flooding or removal of the buoyancy to sink the system to the sea bed slowly, or by winching down the fully buoyant bundle. Both options have obvious difficulties in deep waters.

This technique does present a solution when the bundle configuration is such that mid-depth tow or surface tow is not practical and bottom tow is not suitable due to the nature of the sea bed or pipe-crossings along the route. One specific advantage over other methods of semi-buoyant tow is that weight/buoyancy control is not as critical.

- Mid-depth and off-bottom tow have very similar characteristics and, as such, are to some extent interchangeable. The main advantage of these techniques is that the bundle is towed at a depth below sea level such that the effects of wave action are minimized, but high enough above the sea bed to negotiate rough terrain and pass over existing pipelines.

  With the off-bottom tow technique, as water depth increases the length of the tow cable increases accordingly. This results in reduced directional control. It is under these circumstances that the mid- (or controlled) depth tow technique may be considered more appropriate.
- Bottom tow allows greater directional control, due to the lateral restraint provided by the sea bed against environmental loading. Additional buoyancy need only be used to reduce vessel pull loads, if the submerged bundle weight is excessive. In addition, no special installation procedures are required on arrival at site. The main disadvantage of this method is that, by being dragged across the sea bed, the bundle is susceptible to abrasion and impact damage from sea-bed debris. Use of the bottom-tow method may cause damage to existing subsea installations. Even where measures may be taken to limit damage (i.e. crossing an existing pipeline), the owners of the existing facilities may exhibit a reluctance to expose their facilities to any risk of damage. The selection of this method is therefore very dependent on the route conditions.

As discussed earlier, the selection of towed bundles as a method of pipeline installation will affect certain areas of the pipeline design, such as material selection and thermal insulation selection, where appropriate. Furthermore, if the pipeline design dictates a requirement for exotic materials, such as duplex stainless steel, then its compatibility with other materials used in the fabrication of the bulkheads and carrier pipes must be addressed.

The limitation on the length of the bundles is essentially governed by the length of the site fabrication facilities. Typical single-bundle lengths of up to 5.5 km have been considered. It should also be noted that a number of bundles may be installed in series to increase the overall length of the bundled system. In this way satellite developments in excess of 15 km from the parent facility have been taken to the early stages of detail design.

This multi-bundle technique would generate the requirement for an increased number of bundle termination tie-ins. However, these tie-ins may be readily engineered such that an expedient subsea connection is achieved.

Indeed, similar tie-in techniques may be employed to connect the bundle to the platform and/or the satellite facility.

CONCLUSIONS

In conclusion, selection of the installation method is now an integral part of detailed pipeline design for small-diameter pipelines, and is necessary before the completion of detailed design and specification. Whilst the final selection of the installation method should primarily be based upon economic considerations, it is necessary to develop a detailed knowledge of all of the factors affecting the overall cost of a development which must be considered in detail in conjunction with the pipeline installation costs.

5

The installation of flexible flowlines

W. J. Fraser, Subsea Dynatech Ltd

A BRIEF HISTORY OF THE USE OF FLEXIBLE FLOWLINES

Although flexible flowlines had been used earlier elsewhere with varying levels of success, the advent of their use in the North Sea occurred in the mid-1970s.

Early instances of their utilization in the North Sea occurred in Mobil's Beryl Field (1975), Shell's Brent Field (1976), Beryl (1978), Hamilton's Argyl Field (1979), Beryl (1979) and Chevron's Ninian Field (1979).

These applications were almost invariably concerned with the transport of oil from subsea development wells to a central surface installation for manifolding, processing and export ashore.

At about the same time flexible flowlines were employed extensively in the development of other offshore oil-producing areas, notably in West Africa, Brazil, the Middle East and, to a lesser extent, in the Gulf of Mexico.

SOME ADVANTAGES OF FLEXIBLE FLOWLINES

In certain circumstances the use of flexible flowlines presents useful technical and economic advantages over the use of steel, or rigid,

Advances in Underwater Technology, Ocean Science and Offshore Engineering, Volume II – Underwater Construction: Development and Potential

flowline. In broad terms, these advantages become significant when connections have to be made over relatively short distances, and where the presence of surface installations, or subsea equipment, restricts the access of the laying vessel. The virtue of flexibility is a considerable asset when several flowlines have to be accommodated and connected in close proximity to each other, e.g. at subsea templates and manifolding centres.

The technical benefits may be summarized as follows:

- the ability to accommodate misalignments;
- it may be installed from small, manoeuvrable lay-vessels;
- the ease of installation in restricted areas;
- the ability to curve around obstructions;
- the ability to conform to an irregular sea bed;
- the ease of recovery for repair on the surface.

It is difficult to separate completely the economic from the technical implications, as one is frequently interdependent on the other. However, the principal economic advantages of flexible flowline are derived from the greater ease and simplicity of its installation and connection compared with rigid steel flowline.

In general terms, the economic benefits include:

- the lower day-rate for laying vessel;
- the faster laying speeds attainable;
- the reduced time required to connect ends;
- that span rectification/stabilization are seldom required;
- the flowline may be recovered and used again elsewhere.

TYPICAL APPLICATIONS FOR FLEXIBLE FLOWLINES

As short jumpers (10–50 m)

Situations frequently arise in which two lengths of rigid line have to be connected, and in which there is a limited degree of control over the relative location of the ends of these lines. The effects of thermal expansion may have to be accommodated, or the connection may have to be effected across an area of unstable or irregular sea bed.

Rigid spool pieces are costly, requiring accurate on-site measurement by divers of the relative location and orientation of the flanges to be connected, followed by manufacture and test of sections of rigid line to bridge the measured gap. Self-aligning flanges may have to be incorporated to accommodate tolerances in the measurement and fabrication processes. A large deck area will be required on the surface support facility to fabricate a long rigid spool piece. Slinging and positioning of the completed spool piece will be delicate operations. Exact measurement, fabrication and test of a rigid spool piece must be carried out offshore, where costs are higher and conditions are less easily controlled than in an onshore factory.

Using a flexible jumper to connect sections of rigid pipe, all that need be determined in advance is the approximate location and orientation of the rigid lines to be connected. The flexibility of the jumper can be used to accommodate virtually any misalignment of the pipe ends, unevenness of the sea bed, or obstructions in way of the connection.

Short jumpers can normally be installed in their final location directly by the laying vessel but, should this prove to be impractical, they are sufficiently compact for their position to be accurately aligned by divers working to properly engineered procedures.

As intermediate spool pieces (50–500 m)

In this category of use, operations utilizing rigid steel pipe spool pieces become more difficult, due to the requirement either to handle a single long spool or to fabricate the connection in several spools, which subsequently have to be positioned and connected on the sea bed.

Lengths of less than 50 m of steel pipeline do not achieve the most economic utilization of conventional lay barges. The resulting installation using steel line will tend to be disproportionately expensive.

A spool piece fabricated from flexible flowline may resolve some of the above difficulties. It can be laid as a single length, thus minimizing the number of connections (and the number of potential points of leakage) in the final installation. Misalignments between the flanges to be connected are readily accommodated. The vessel required to install the flexible spool piece will usually be smaller (and therefore more manoeuvrable and less expensive) than that required to achieve a rigid connection.

When a flexible flowline is used to connect flanges across a distance of more than 50 m, the length and weight of the resulting spool piece is greater than can be easily worked by divers. It therefore becomes more important to specify a sea-bed track which will enable the laying vessel to position the majority of the length of the spool piece accurately. Divers may then be employed to align and connect the ends.

As production flowlines (500–15 000 m)

For connections over substantial distances, valid arguments for using flexible flowline still exist. These arguments are primarily based on the simpler and more-straightforward installation procedures possible, but the naturally higher cost of the flexible, compared with the rigid, product becomes significant.

The views of proponents of flexible and rigid steel flowline may be found to differ somewhat on the point at which one type of line becomes more competitive than the other. However, it would be fairly safe to conclude that, based on the current market, and for continuous straight installations, flexible flowline would not be competitive over greater distances than those set out below:

*Line diameter (in.)**	*Distance (m)*
4	15 000
6	10 000
8	6 500
10	2 500

Due to restrictions on the volume and weight of continuous lengths of flexible flowline which may be handled during fabrication, long installations would have to consist of several sections of line. The maximum continuous length in which flexible flowline is readily available for typical production pressure ratings is given below:

Line diameter (in.)	*Maximum section length (m)*
4	3 500
6	1 500
8	1 000
10	600

Provided that the laying vessel is adequately equipped, several reels of flexible line may be mobilized simultaneously. As each section is laid, the next section is connected to it, and the entire length may thus be laid as a continuous operation. This type of operation may have certain advantages during variable weather, as there is the opportunity to temporarily abandon the laying operation between sections should the weather blow up, and resume the operations when conditions improve.

*1 in. ≈ 2.54 cm.

As flexible risers (100–500 m)

Recent early production developments have seen the introduction of the use of flexible flowline for production risers between fixed installations on the sea bed and floating platforms. Typically, such developments deploy the flexible line in a soft catenary over the edge of the floating platform towards the subsea facility. Midwater buoyancy may be attached to the flexible to achieve a compound catenary in an S-shaped configuration.

Claimed benefits of such systems include:

- simplified overall system design;
- ease of installation;
- ease of disconnection, when required;
- free access through the moonpool for well workover, etc.

TYPICAL CONSTRUCTION OF FLEXIBLE FLOWLINE

An understanding of the construction and characteristics of a flexible flowline is useful, if not essential, to the preparation of correctly engineered applications and installation procedures.

Typically, flexible flowline consists of several layers, each of which performs a specific function in meeting the service requirements of the line.

The innermost component is normally a flexible tube formed from a close-pitch, spirally-wound continuous steel strip, each turn of which is interlocked mechanically with the previous turn. This inner metallic liner resists external forces and pressures which might tend to collapse the line, and provides a calibrated bore for the finished flowline. It also provides an internal surface which resists erosion by abrasive particles entrained in the fluids carried, and withstands the scuffing effect of pigs and through-the-flowline tools which may be required to service the installation.

Around this innermost layer an impervious continuous layer of flexible polymer, normally plastic or rubber, is extruded. This extruded layer forms the containment for the fluids to be carried when the line is in service. Unsupported, it would be totally incapable of withstanding the internal pressure in the line – rather like the inner tube of a tyre.

The third layer of a typical flexible flowline consists of another close-pitch spirally-wound flexible steel casing. The function of this layer is to provide a continuous supporting surface around the polymer containment, to prevent this polymer from being extruded through subsequent layers of armouring under the effect of internal line pressure. Similarly to the innermost layer, each turn of this third layer is interlocked to the preceding and succeeding turns.

The fourth layer (which may be omitted in lines designed for lower service pressures) consists of a spirally-wound layer of rectangular wire, sized to provide containment of the internal pressure within the flowline when in service. In order to contain efficiently the hoop stresses in the line generated by the internal pressure, this fourth layer is wound with a relatively close pitch, and is opposite-handed to the third layer beneath it.

The fifth and sixth layers together comprise that part of the line which supports the axial tensions applied to the line. These tensions are created, during installation, by the weight of the catenary suspended from the laying vessel and, in service, by the effect of internal pressure acting on the ends of the line. The layers are formed from multiple thin flat bars, contra-wound in two separate layers with a relatively long pitch around the pressure containment. The pitches of the fifth and sixth layers are calculated such that a balanced construction of the line is achieved, i.e. when under the effect of internal service pressure the Bourdon effect is eliminated, and there is no tendency for curves or bends in the line to straighten out.

The seventh (and final) layer consists of a polymer extrusion around the finished construction. This layer serves to maintain the composition of the previous layers, and prevents them from becoming displaced, or slackened, with respect to each other. It also provides mechanical protection of the line from impact and abrasion during installation and service, and protects the line from the corrosive effects of open contact with sea water.

The construction of some typical flexible flowlines is illustrated in Fig. 1.

Fig. 1 Typical construction of various sizes of flexible flowline.

For specialized applications additional components, such as thermal insulation, trace heating elements, etc., may be incorporated at appropriate points in the construction during fabrication of the line. Flexible risers are continually working in a dynamic environment, and are therefore more exposed to the effects of internal friction between the components of the line than are flexibles for predominantly static use. To minimize the effects of this internal friction, and eliminate wear between adjacent layers of steel windings, an intermediate sleeve of a polymer material may be interposed between them.

To complete each section of flexible line, end-fittings are installed. Although these may be to virtually any specified pattern (fixed flange, clamping hub, rotating flange, etc.), all are attached to the flowline in a similar way. The procedure is complex, as sealing against the effects of internal and external pressure, maintenance of electrical continuity, and anchoring of all the constituents of the reinforcing windings must all be reliably achieved. As these end-fittings effectively solidarize all the individual constituent elements of the line, full flexibility is not developed until some 3–5 m distance from the end-fitting. This has important implications during the design of acceptable system layouts, and during operations to install and connect the line.

The physical characteristics of some of the line sizes and pressure ratings in which flexible flowline is available are given in Table I.

Table I
Physical characteristics of some available flexible flowlines

i.d.		*Working pressure*		*o.d.*		*weight in air*		*Minimum bending radius*	
in.	*mm*	*psi*	*bar*	*in.*	*mm*	*lb/ft*	*kg/m*	*ft*	*m*
1	25.4	3 000	207	2¼	57.5	4.4	6.6	1.22	0.37
1	50.8	3 000	207	3¼	81.5	7.2	10.7	1.75	0.55
2	50.8	10 000	690	4¼	108	20.7	31	2.3	0.7
2½	63.5	2 000	138	4	101.5	9.7	14.4	2.2	0.66
3	76.2	5 000	345	5	125	21	31.5	2.7	0.81
3	76.2	10 000	690	$5\frac{3}{16}$	131	24.5	36.5	2.8	0.85
4	101.6	10 000	690	7	178	51.7	77.2	3.8	1.16
4	101.6	3 000	207	$5\frac{9}{16}$	141	14.6	21.8	3	0.91
5	127	5 000	345	7¼	183.5	38.5	57.5	4	1.2
6	152.4	2 000	138	7¾	195.5	27.7	41.3	4.2	1.3
6	152.4	7 500	517	9	229	73.3	109.4	4.9	1.5
8	203.2	2 000	138	$10\frac{3}{16}$	257	38.1	56.9	5.5	1.7
8	203.2	5 000	345	11½	292	99.5	148.5	6.3	2
10	254	1 500	103	12¼	310	55.5	82.5	6.6	2.1
10	254	4 500	310	13⅞	353	139	208	7.5	2.3
12	304.8	1 000	69	14½	368	72.1	108	7.9	2.4
12	304.8	2 500	172	15¼	388	106	158	8.3	2.55
16	406.4	750	52	20	504.5	162	242	10.8	3.3

THEORETICAL BEHAVIOUR OF FLEXIBLE FLOWLINE

Flexible flowline, although possessing some internal stiffness to bending, is relatively heavy. When suspended between the laying vessel and the sea bed the size of the resulting catenary is sufficiently large, in relation to the line diameter, to enable the effects of line stiffness to be ignored for the purposes of calculation, and the line may be treated as a freely-bending heavy flexible.

As a general guide flexible flowline can support substantial axial tension, but has minimal strength in bending. Installation procedures must take this into account, ensuring that all manoeuvres can be performed with the line in linear suspension.

The forces acting on such a freely-bending heavy flexible, suspended in catenary, are shown in Fig. 2. Its behaviour is predicted by the relationship:

$$y = c\left[\tfrac{1}{2}(e^{x/c} + e^{-x/c}) - 1\right]$$

where y = height of the catenary,
x = horizontal extension of the catenary and
$c = H/w$, where H = horizontal component of tension and
w = unit weight of the flexible.

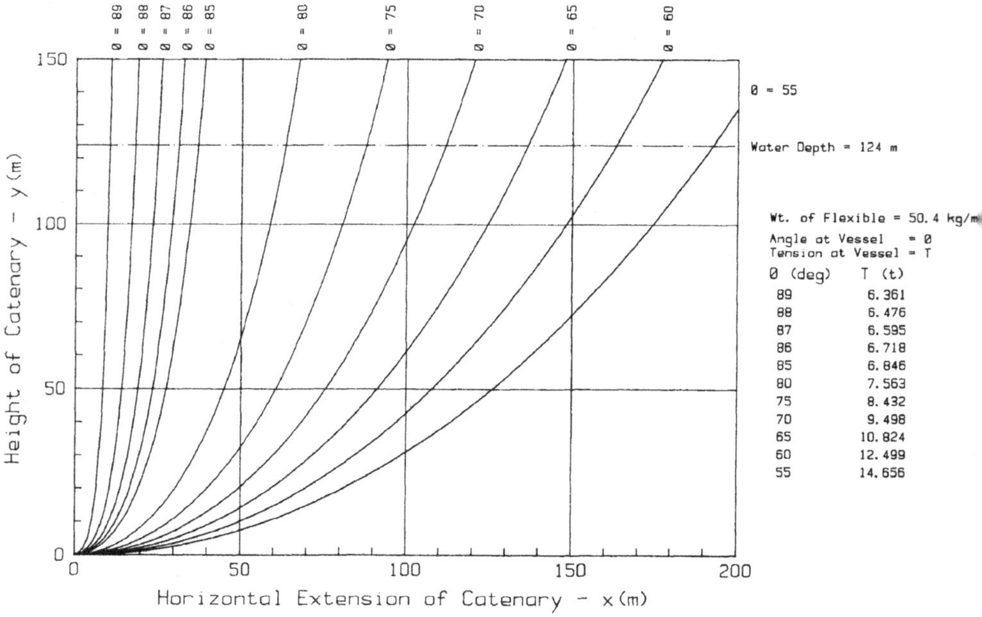

Fig. 2 Forces acting on an element of flexible suspended in catenary. *H* and *V* are the horizontal and vertical components, respectively, of *T*.

From this basic relationship, a series of equations may be developed interrelating the following parameters of the catenary:

w – unit weight of the flexible;
x – horizontal extension of the catenary;
y – height of the catenary;
s – length of flexible in suspension;
θ – angle of upper end of catenary to the horizontal;
T – tension in the flexible.

Knowledge of any three of the above parameters enables the remaining three to be predicted mathematically.

The unit weight of the flexible (w) is available from the manufacturers of the flexible. The height of the catenary (y) can be readily calculated as the sum of the water depth and the deck height of the laying vessel. The angle of the upper end of the catenary (θ) or, less readily, the tension in the flexible (T), may be measured on board the laying vessel. It is therefore possible to predict mathematically, with confidence, the shape of the catenary suspended in the water, and the distance of the contact point on the sea bed away from the laying vessel.

Buckling of the flexible does not present a problem if the line is correctly handled. When the line is suspended linearly, and subjected only to smooth and gentle movements, its inherent stiffness is adequate to ensure that it does not adopt curves of less than its minimum bend radius when being laid on the sea bed under the sole action of its own weight. At higher points in the catenary, where the line tension is considerably greater, care must be taken to ensure that curves are maintained above a safe radius. Bend restrictors are frequently employed for this purpose.

Catenaries that might occur during a typical laying operation in the North Sea are presented in Fig. 3, together with corresponding values of the angle of the catenary to the horizontal at the vessel (θ) and the tension in the flexible at the vessel (T).

FACILITIES REQUIRED FOR INSTALLATION OF FLEXIBLE FLOWLINE

The most practicable method of transporting flexible flowline is by winding it on to shipping reels. This ensures that the minimum bending radius of the line is correctly maintained throughout the shipping operation. Transport reels are normally configured so that they may be rotated either by driving the hubs in an appropriate

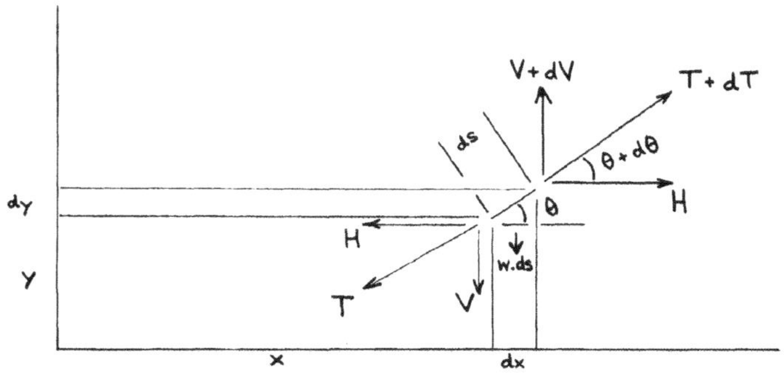

Fig. 3 Physical parameters of catenaries for a typical North Sea flexible flowline installation.

winch or by rolling the rims of the reel on a suitable dolly base.

The maximum weight of a typical transport reel fully laden with flexible flowline is 200 tonnes. Heavy and expensive craneage may be required to handle reels of line during loading and unloading operations. It may therefore be preferable to carry out loading of the laying vessel alongside the manufacturing plant, where the line may be spooled directly on to the laying equipment on board. It is inadvisable to attempt to transfer full reels of line by crane at sea, although transpooling of the line between vessels may be practicable.

Some form of winch is essential for installation operations, so that the line may be lowered to the sea bed in a finely controlled manner and so that, should the requirement arise, line may also be recovered back on board the laying vessel. As there is always some friction present in the system, this implies that the winch should be capable of producing a minimum of 125–150% of the maximum calculated line tension at the greatest drum diameter at which this may occur.

A wide range of different laying systems have now been evolved. This range comprises:

- independent winches with fixed reels;
- independent winches with interchangeable reels;
- mobile winch systems which can be moved along the deck of the laying vessel to pick up and rotate any one of a number of different reels;
- specialized vessels on which large quantities of continuous flexible line are stored in rotating baskets and line tension is controlled by multiple-caterpillar line tensioners;
- different combinations of the above systems.

Regular and even layering of the flexible line on the transport reels is essential, both to safeguard the line itself and to obtain efficient packing of the line within the available volume on the reel. To maintain full control over the line during its installation, level winding equipment is also essential on board the laying vessel.

Care should be taken to ensure that heavy end-fittings are adequately padded to prevent them causing damage to the external surface of adjacent turns of the line whilst the reels are in transit.

The laying vessel itself should have the following features:

- adequate stability for the required deck load;
- adequate deck reinforcement for heavy localized loadings;
- dynamic positioning system for holding station about a set point, or navigating accurately along a defined track;
- adequate deck space to leave 20–30 m of free deck area between laying winch/level winder and stern;
- laying roller or spout to maintain a safe bend radius in the line as it passes over the stern of the vessel;
- adequate deck craneage – particularly over the stern – to facilitate handling of the line on deck.

Many of the larger dynamically positioned supply-type vessels provide the required facilities for use as laying vessels, although the maximum amount of flexible flowline which may be embarked at one time is generally less than 800 tonnes. For larger amounts of line the more recent semi-submersible multi-service vessels may provide greater deck space and load-carrying capacity. Specialized laying vessels have been developed by the manufacturers of flexible flowline, which are capable of embarking and continuously laying more than 1000 tonnes of line on each trip.

Testing of the line is normally specified to be carried out following its installation on the sea bed. Frequently it is possible to have this testing conducted from the fixed surface installation to which the flexible flowline has been connected, thus achieving a complete test of the connected system. This is usually the most cost-effective and expeditious method of doing so. Alternatively, for longer lengths of flexible flowline testing may be carried out from the deck of the laying vessel, before laying down the final end of the line on to the sea bed. To reduce the time required to fill the line and bring it up to test pressure, a high-pressure, high-volume, positive-displacement pump is essential.

INSTALLATION METHODS

For longer lengths of flexible flowline the optimum geometries for laying catenaries have been largely determined from experience. For laying flexible flowline in a straight line, at some distance from curves in the track, an angle of 80–85° to the horizontal would normally be appropriate at the departure point from the laying vessel. The resulting catenary ensures that a sufficient horizontal component of tension is present in the flexible to eliminate the effects of minor variations in the course of the laying vessel along the track in which the flowline has been laid. However, when close to curves in the track, or when in the vicinity of terminations, a more vertical catenary with an angle of 87–89° is appropriate. The corresponding reduction in the horizontal component of tension in the catenary then ensures that line tension is reduced, and the line will not be dragged out of the curves in the track into which it has been laid.

Having determined the appropriate geometry of the laying catenary for the section of track to be laid, the course along which the laying vessel must be navigated is calculated, together with the corresponding amounts of flowline which must be paid out, to ensure that the flowline contacts the sea bed along the required track and, most importantly, remains there. When initiating or terminating a section, or when negotiating curves in the track, the geometry of the catenary may need to be altered. This is easily done by simply paying out or reeling in a calculated amount of the flowline, or by moving the vessel a calculated distance, or by a combination of both of these measures.

The influence of tidal currents running transversely to the catenary will tend to displace the touchdown point to one or other side of its theoretically calculated position. However, it is difficult to make accurate tidal measurements over the height of the catenary. Operational experience in the North Sea indicates that the effect of tidal currents is small compared with the effect of weight, and is, for most practical purposes, negligible. Experience indicates that a positional accuracy of the installed line of 5% of water depth is readily achievable, with figures of 3% or less frequently being attained.

When installing flexible spool pieces whose length is less than the water depth, or when conducting J-tube pull-in operations, the geometry of the resulting catenary is complicated by the presence of the lighter pull-in and lay-down cables, and the concentrated point loads formed by the end-fittings at the ends of the flexible flowline. The compound catenaries involved must be fully analyzed in order to maintain the accuracy of the installation, and to ensure that the correct angle of entry is maintained at the J-tube throughout the pull-in operation.

Subsea flowlines are normally buried, to provide protection from trawling gear, anchors and other potential causes of fouling or damage. Flexible flowlines, by their nature, lend themselves readily to trenching during laying operations. A mole-type of plough, or trench cutter, is normally used, deployed and controlled from the laying vessel. During operations to position or manoeuvre the trenching machine on the sea bed, the flexible flowline may be maintained clear of the machine by picking up on the catenary, and similarly relocated in the machine by lowering down on the catenary before recommencing laying and trenching operations.

A typical configuration of equipment for laying and simultaneously trenching flexible flowline is shown in Fig. 4.

Fig. 4 Typical configuration of equipment for simultaneous laying and trenching of flexible flowline.

MISCELLANEOUS COMMENTS

Installations completed to date have included the simultaneous laying of two flexible flowlines and two control umbilicals, and J-tube pull-in of flexible flowline at both initiation and completion of lay.

Flexible flowline is particularly well-suited to taking advantage of the many designs of remote, diverless connection system which are currently under investigation for use in deeper waters.

Techniques for recovering flexible flowline to the surface for repair following accidental damage are well advanced, and reduce the overall system downtime associated with the hyperbaric welding techniques normally required to repair rigid steel lines.

Some 6.5 km of flexible flowline has recently been recovered from a water depth of 120 m, following 7 years of virtually continuous oil and gas production service. The line was in very good overall condition, and successfully completed detailed inspection, pressure testing, and re-certification for further subsea service elsewhere. This is an obviously significant development for future temporary production schemes.

Although there are inevitably certain drawbacks associated with the use of flexible flowline, it now has a proven track record of more than 10 years of service in environments ranging from the North Sea to the Tropics. It may be expected to continue to play an important role, particularly in the realization of early production developments, for many years to come.

ACKNOWLEDGEMENTS

The provision of illustrative material by Coflexip s.a., and their permission to reproduce it in this paper, are gratefully acknowledged.

6

Subsea construction from monohull vessels

T. P. Littlefield, BUE Services Ltd

INTRODUCTION

Monohull vessels make up by far the most widely used hull form on the waters of the world, and not least in the field of subsea construction. The workhorses of the offshore industry, the monohulls, have been designed, modified and often converted to facilitate a multitude of tasks.

It is acknowledged that the semi-submersible hull form is playing an increasing role in subsea construction, but it is the numerically superior and often more versatile monohulls upon which so much sttill depends. It is not the purpose of this presentation to analyse comparisons, but rather to illustrate the abilities of monohull vessels in subsea construction.

DESIGN AND SPECIFICATION

Subsea construction vessels tend to fall into two categories. The first are those which are designed and constructed to perform a task or tasks, with diving very much a secondary capability. The second category consists of vessels that have a major diving capability but are also able to carry out a number of other tasks.

Advances in Underwater Technology, Ocean Science and Offshore Engineering, Volume II – Underwater Construction: Development and Potential

Vessels forming the former group are the crane barges, lay barges and general-purpose construction vessels; for example, Reel Ship *Apache*, and the newly built DB *Challenger*. The latter group are the diving support vessels, for example *Arctic Seal* and newly built *Seawell One*.

Both categories must be able to perform to meet high specifications and the following considerations must be made during design and building:

- dimensions;
- power;
- positioning systems;
- subsea construction facilities.

Dimensions

Length, beam and draft make up the vital statistics which dictate the actual size of a vessel. However, it is the way in which these measurements are arranged that is important.

The monohull must be full enough in the beam to provide a stable and sea-kindly work-platform, but must be an efficient profile to enable good transit speeds to be achieved and economically maintained. The draft of a vessel must be considered, not only as a component of stability but also in order to keep a wide choice of ports and harbours available, as one of a monohulls strengths is its ability to go alongside a jetty or quay for load-out and mobilization.

When considering a vessel for specific workscopes, size plays an important part in the selection process. For the most part, larger vessels will be utilized for the heavier tasks and when large numbers of personnel are required. This does not by any means preclude the smaller vessels, as they come into their own for lightweight tasks and where access may be restricted. Often a large vessel may carry out a single part of a project and a smaller vessel may then take its place to complete the work.

Deck and working areas are proportional to the overall dimensions of a vessel and the accommodation requirement. The scope of work will dictate how much deck is required, and therefore vessel selection. The loading capability of a deck must also be considered, as specialist construction equipment may cause problems requiring the fabrication of additional grillage and sea fastenings.

Personnel accommodation is critical, and a vessel can only carry out work for which she can accommodate and victual the necessary crew. If a vessel is short of accommodation but is working in the vicinity of another installation, it maybe possible to shuttle personnel. This is, however, inefficient and requires accurate planning and good conditions to ensure personnel are in the right place at the right time.

Power

Monohull subsea construction vessels must have abundant available power. In today's market the emphasis is on self-propelled vessels with dynamic positioning. This ensures flexibility for working, and speed during relocation, but requires considerable ability to provide power for main propulsion, position keeping and operating equipment.

A vessel must have the ability to remain on station in all but the worst conditions, and although this is partly a function of hull size and design, powerful main propulsion and thrusters are required. With this in mind, the designers must also consider economy and achieve a balance with fuel consumption during normal conditions.

The intervals between bunkering must be optimized, and generally designers are able to match-up the consumption of fuel oil, victuals and other consumables to make replenishment more efficient, either if the vessel remains on location or returns to port.

Positioning systems

The ability for a subsea construction vessel to take up and maintain a position is critical. There are essentially two methods:

- Mooring;
- dynamic positioning.

Mooring involves the deployment and tensioning of either anchors on the sea bed or mooring lines secured to an existing structure. Mooring still has applications, for example if a vessel is programmed to remain on a location for a long period or in locations where water depths make for rapid anchor handling. For the most part, however, mooring systems are used only to back up the sophisticated dynamic positioning equipment which has become so universal.

Dynamic positioning (DP) has developed considerably over the past 15 years and has now become standard in all but a few types of subsea construction vessel. By using a number of sources of information the DP computers can automatically call on the directional thruster systems. Sources include long, medium and short range radio signals transmitted from satellites, shore stations or local installations. Acoustic signals transmitted from transponders installed on subsea structures or deployed by the vessel herself are used for relative positioning. The third type of source is mechanical and includes tautwire. This system involves the deployment of a weight to the sea bed and, by using compensators, the wire tension remains constant. Sensors then inform the computer of changes in relative angles and directions of the wire, which are then assessed as position changes. The flow of information is continuous, which in turn keeps the system constantly reacting to maintain the vessel in the desired location, and a vessel's ability to maintain station is dependent on the speed of reaction and available power in the DP system.

In order to allow good margins of safety for DP operations (especially when carrying out diving operations) a series of specifications must be observed. These include the requirement for full redundancy of computers and sensors, equipment-failure indicators and power-level warning systems.

Subsea construction facilities

The equipment outfit of a monohull subsea construction vessel will be very much dictated by dimensions and power. However, many general-purpose vessels can be adapted and prepared for specific tasks by making temporary modifications.

When considering a vessel for subsea construction tasks the following facilities will be reviewed:

- craneage: to include cranes, A-frames, derricks and davits, their locations, lifting capacities and reach;
- diving systems: saturation, surface-orientated and ROV;
- workshop and fabrication: for maintenance, structural and pipeline construction;
- welding: surface and subsea;
- cargo and stores: dependent on sizes, but the type of cargo is an important consideration;
- helideck: aircraft size and type, and refuelling capability;
- emergency intervention: fire-fighting, oil-spill management rescue and hospital facilities.

BALMORAL FIELD DEVELOPMENT PROJECT

The recent North Sea Sun Oil Balmoral Field Development Project provides an illustration of the varied capabilities of monohull vessels in subsea construction, and a brief review of their activities during this successful project follows.

The Balmoral Field, with a water depth of 148 m, was classified as marginal, and the selected method of production was a central template with satellite wells and associated floating production vessel.

Offshore development programme

In June 1984 Heerema's SSCV *Hermod* installed the central template and, shortly afterwards, the *Western Pacesetter 1* began the drilling programme. Drilling at the template continued throughout the winter of 1984–85 until the end of May 1985, when the rig moved off, thereby allowing access to the template and the main subsea construction programme to begin.

The outline workscope included the following major items.

Santa Fe Reel Ship Apache

(1) Lay six 6 in.* water-injection infield flowlines.
(2) Lay three 4 in. production flowlines, piggy-backed with three 2 in. production annulus flowlines.
(3) Lay the 14 in. oil export pipeline from the template to the Brae-Forties trunkline.
(4) Transport to field and lower three flowline bundles between the template and gravity riser bases.

ITM Offshore DSV Installer

(1) Carry out "as laid" surveys of all pipeline laid by *Apache* using Scorpio ROV.
(2) Trench all pipeline laid by *Apache* using British Telecom's Seadog trenching vehicle.
(3) Carry out "as trenched" survey of all pipelines laid by *Apache*.

Heerema Marine Contractors DSV Explorer II (on charter to BUE Services)

(1) Transportation and installation of four gravity riser bases (PLEMS).
(2) Inhibition, underwater installation, tie-in and pressure testing of 30 flexible spool pieces.

*in. ≈ 2.54 cm.

(3) Pigging, inhibition and pressure testing of 12 flowlines.
(4) Measurement for fabrication and subsequent installation of a 14 in. diameter hard pipe spool piece for the oil export pipeline.
(5) Installation of three piled dead-man anchors for pipeline initiation.
(6) General diving support of Santa Fe *Apache*.
(7) Correction of freespans on the pipelines by installing grout-filled supports.

HAM dredging: Dredge and Fall Pipe Vessel HAM 310

(1) Correction of freespans on the pipelines by gravel dumping.

Safe Offshore MSV Karinia (on charter to Wharton Williams)

(1) Installation of the 14 in. oil export pipeline tie-in and bypass facility at the Brae-Forties trunk line.
(2) Installation of manifolds at the subsea template.

On completion of the 1985 construction programme, drilling and well completions continued at the template and satellite wells. This programme ran on throughout the winter of 1985–86 until the end of July, when the drilling rig moved off the template and the second major construction phase commenced.

Oddberg Bergship DSV Arctic Seal (on charter to BUE Services)

(1) Inhibition, underwater installation and tie-in of 12 flexible spool pieces to the satellite wells.
(2) Position and connect the subsea control umbilicals at the satellite wells.
(3) Monitor and carry out leak checks during hydrostatic pressure and tree functions testing.
(4) Debris clearance and recovery of driven anchor pile from vicinity of template.

Norcem DSV Norskald (on charter to ITM)

(1) Initiate and lay nine control umbilicals from the subsea template to the satellite wells.
(2) Carry out "as laid" surveys of each control umbilical.

Safe Offshore MSV Karinia (on charter to Wharton Williams)

(1) Provide support for test platform during template and flowline hydrostatic testing.
(2) Assist initiation and connect nine control umbilicals at the template.
(3) Install control pods and support control umbilical function testing.
(4) Carry out a variety of tasks to prepare the subsea template for the arrival of the floating production vessel.
(5) Make weld repair on 6 in. flowline in template vicinity.
(6) Install and tie-in the flexible risers between the gravity riser bases and the floating production platform.
(7) Commission the bypass facility at the Brae-Forties trunkline.

Allseas Lorelay

(1) Installation of piled anchors for floating production platform mooring system.

Having listed the vessels and summarized the workscopes carried out during the 1985 and 1986 North Sea Sun Oil Balmoral Development Project, it is now proposed to take a closer look at the Monohull Construction Vessels and the way in which they carried out their tasks.

THE MONOHULLS AT WORK

Reel Ship Apache

The unique abilities of the *Apache* made her ideally suited to the task of laying a variety of lengths of different diameter steel pipelines.

A total of 17.0 km of 6 in. pipeline was laid between the central subsea template and the six satellite water-injection wells. This was followed by dual lay of 4 and 2 in. production and annulus pipelines from the template to the three satellite production wells – a total of 15.0 km. The 14 in. oil export line was laid in two lengths, giving a total of 14.4 km between the template and the tie-in to the Brae-Forties trunkline.

By utilizing this method to lay steel pipe, environmental conditions were assured for welding and fabrication, as this was carried out onshore at the Evanton spooling base and commenced well before the initial mobilization. During laying operations, pipeline welding continued onshore in preparation for spooling. The only fabrication tasks required offshore were those of anode attachment and flange weldment before laydown.

During the Balmoral Project, the return trips made by the *Apache* were used to transport the three riser flowline bundles to the field. On each occasion the riser flowlines, held in supporting frames, were secured to the davits located on the portside. The procedure for laying the bundles on the sea bed was as follows.

The *Apache* was positioned on DP over the touchdown location and, using a sea-bed acoustic transponder array, with transponders at each end of the riser flowlines, landing of each bundle between the template and the gravity riser bases was achieved with an accuracy of better than ± 1.0 m. Once the flowlines were landed, the davit wires were slacked off to allow *Apache* to move off to starboard. This enabled DSV *Explorer II* to set up on DP and deploy a diver to release the lowering wires from the supporting frames, thereby enabling *Apache* to continue pipelaying.

DSV Installer

The *Installer* provided survey and trenching services in support of *Apache* pipelaying and, although equipped with an inbuilt saturation diving system, this was not a facility required to complete her tasks. Instead, her inherent diving support vessel abilities were utilized to provide accurate positioning, and a stable platform to operate the Seadog trenching vehicle and Scorpio ROV.

DSV Explorer II

Explorer II was constructed to fulfil a primary role as trenching support vessel equipped to operate the "Eager Beaver" trenching vehicle. However, many design features made her a near perfect choice for the varied workscope of the Balmoral Project, and a closer examination illustrates her abilities as a monohull vessel for subsea construction.

The task which *Explorer II* was particularly suited to was the transportation and installation of the four gravity riser bases. These PLEMS had air weights which ranged from 65 to 90 tonnes, and it was possible to transport only two at a time. The vessels unique 120 tonne heave-compensated A-frame made overboarding and installation on the sea bed possible to an accuracy of ± 0.5 m – well within the specifications. By using rigid rigging it was possible to carry out the touchdown and slack the lift wire in one operation without dropping the swivel on to the riser base. Positioning during installation was by acoustic transponder and Micronav system, with a series of position fixes combined with bathymetric readings taken before a diver released the rigging.

In order to initiate pipelaying in the vicinity of the subsea template it was necessary to install three swivelling anchor sheaves which were installed by *Explorer II* using the Anchortech piling hammer.

The swivel anchor sheave assemblies had previously been set on to the piles and for the first pile, the hydraulic subsea hammer had also been fitted. The whole rig was overboarded using the vessel's crane, and was lowered to the sea bed, final touchdown positioning being achieved using the template acoustic transponder array. Once on the sea bed, the initial pile penetration was made by the rig's own weight, and the pile was then driven using the hammer operated from *Explorer*'s deck. After piling to depth, a diver was deployed to derig the hammer for recovery. During the first swivel anchor sheave installation a number of minor problems were encountered, attributable mainly to the poor soil conditions. For the second and third anchor piles, small modifications were incorporated which resulted in perfect installations.

Before pipeline initiation from a swivel sheave anchor, *Explorer II* set up on DP at the template location and deployed a diver to the sheave. The end of a 15 mm wire was lowered to the diver, who rove the wire through the locking gates and sheaves of the anchor. Once rove, the wire was recovered to surface using the crane. With this messenger wire rigged, the SDC was recovered and the *Explorer II* then manoeuvred to lay out the wire on the sea bed and buoy-off each end.

The *Explorer II* then stood off from the immediate vicinity of the anchor to enable *Apache* to move in and recover the two buoys. The deck crew onboard *Apache* then connected the A & R (Abandon and Recovery) wire and winched-in the messenger wire. By paying-out the A & R wire and taking up on the messenger, the 60 mm A & R wire was pulled through the sheaves of the dead-man anchor and recovered to deck. This end was then attached to the pulling head flanged to the pipeline, and the *Apache* began overboarding the pipeline and pulling-in the A & R wire. The conical pulling head was then pulled into the swivel anchor and the latching bars held it in position.

The *Apache* slacked off the A & R wire and laid away to a position enabling *Explorer II* to set up and give access to her divers. After confirmation that tensions were minimal on both pipe and wire, the diver fitted a hydraulic cutter and cut the wire, enabling pipelaying to continue and *Explorer II* to resume her construction tasks.

On completion of each pipelay the *Explorer II* was tasked to pig, gauge and inhibit the lines. This was required before each line was trenched, and involved diving support for the Chorley Engineering spread.

Upon confirmation that a pipeline had been abandoned by *Apache*, *Explorer II* set up at the initiation laydown head to enable a diver to prepare the pig receiver. *Explorer II* then relocated to the laydown head and deployed the fill and monitoring hoses. These were then connected to the pig launcher by the diver who, upon instruction from the test engineer, operated the launcher release pins and fill valves.

On completion of pig running, the diver derigged the fill hoses and *Explorer II* relocated to the pig receiver and deployed the SDC. The diver removed the pig receiver and, using oxy-arc cutting equipment, disconnected the conical pulling head. The pig receiver was then recovered to the surface using the crane and the pigs were removed. This confirmed that all pigs had transitted the pipeline and enabled inspected of the gauging plate. *Explorer II* then return to the outfield end of the pipeline and deployed a diver who removed the pig launcher and replaced it with a blind test-flange which was installed using hydraulic flange tensioning tools.

During the 1985 programme *Explorer II* and her diving crew installed and connected 30 flexible spool pieces to tie-in the template and riser bases with the satellite well flowlines and the riser flowlines. The spool pieces were manufactured by Dunlop Oil and Marine, and varied in diameter between 2 and 10 in. and were from 10 to 35 m in length. The spool pieces were transported to the field onboard *Explorer II*, and required three load-outs. The first and second were combined with loading-out the riser bases which took place at Leith, and the third at Aberdeen.

Overboarding and installing the flexible spool pieces required special procedures as follows.

Explorer II set-up on DP and deployed the SDC to the sea bed, where the diver locked out, located the template flange and removed the blind flange. While this operation was being carried out subsea, the appropriate spool piece was prepared for installation. Preparation included flooding with inhibitor, attaching the Seamark identity labels and securing tie-back tag lines. The spool piece was then overboarded by either lifting directly with the crane or, in the case of spool pieces more than 25 m long, manoeuvred over the stern through a chute and then transferred to the crane. Once overboard, the spool piece was lowered to a position 5 m above the tie-in flange. With directions from the diver and ROV the spool piece was lowered further to allow the diver to secure the two tie-back tag lines to the structure. The lay-down of the spoolpiece then commenced and, by manoeuvring the vessel and crane boom as the load was lowered, the spool piece was laid-down in the desired configuration with the flanges in position for the diver to complete the tie-in.

On completion of lay-down, the diver disconnected the lifting sling and tag lines and commenced the flange connections. Techniques for the tie-ins varied with each location, but followed the following basic procedure.

A downline from a 2 tonne constant-tension air-winch was lowered to the worksite and secured to the pipeline using a nylon sling. The pipeline flange was then lifted 1.5 m off the sea bed to enable the diver to place a supporting trestle beneath it. The pipeline was then lowered on to the trestle and the lift wire disconnected. The diver then secured two 0.5 tonne lifting bags to the spool piece. By inflating the lifting bags the spool piece was raised above the sea bed, approximately in line with the pipeline flange. The diver then installed two Hydratight wire rope pullers between the flanges and "sucked" the spool piece towards the pipeline. When the flanges were approximately 50 mm apart the diver inserted the BX ring seal gasket into the flange groove. The flanges were then closed sufficiently to ensure that the ring seal was in position and the studbolts could be inserted. With all studbolts located and spanner tight, the diver then assembled the flange tensioning tools and connected the hydraulic hotline. After testing the tools, the pressure was increased in stages up to the predetermined pressure required to seal the flanges and, at each pressure stage, the diver took up slack by turning up the inside nut on each studbolt, thereby retaining the tension on the flanges.

On completion of making-up the flange connection the divers derigged the pipeline and spool piece lifting equipment, and relocated to the other end of the spool piece to complete the connection.

In addition to the flexible spool pieces the *Explorer II* and her construction crew installed the rigid spool piece to tie-in the 14 in. oil export line. This task began with the tight-wire measurement between the template flange and the pipeline flange.

The master and slave units were fitted to the template and pipeline flanges, and with the piano wire drawn tight the diver read off the required angles for surface recording. The diver then locked the units and marked the wire before derigging. The tight wire units were then recovered to deck and transferred to the *Apache* for fabrication of the spool piece. The overall workscope of the *Explorer II* and the requirements for *Apache* to return to Evanton to spool the second length of 14 in. pipe meant that the spool piece was fabricated with the convenience of shore facilities. Upon her return to the field, *Apache* overboarded the spool piece and landed it on clear sea bed. Divers deployed from *Explorer* detached the lift wire and transferred the load to her crane. Acoustic transponders were fitted to each spool piece flange and, with visual assistance from diver and ROV, it was relifted and landed in position for connection using handling frames and hydraulic bolt-tensioning equipment.

Explorer II, as well as supporting the subsea construction programme, also provided the hydrostatic testing spread. All tied-in pipelines were required to complete a 24 h hydrostatic pressure test and all tied-in riser flowlines were required to complete a 6 h leak test.

The hydrostatic testing was initiated at the subsea template with *Explorer II* set-up over the location. The test hoses (up to three sets at a time) were deployed over the vessel's starboard side and all connections made to test ports in the template pipework. Due to the relative proximity of the test ports and the construction tasks at the template, the divers were able to continue making up spool piece connections while tests were running on adjacent pipelines.

During the "as-laid" surveys of the pipelines, a number of freespans were located which were outside the required specification. Remedial action was necessary and *Explorer II* took on a grouting spread and plastic rock-guard material.

The *Explorer II* set-up at the freespan locations and deployed the ROV and diver to measure heights and lengths. At each freespan location the divers wrapped the pipelines with rock-guard material and, in addition, installed grout-filled support pyramids. The final operation to correct the freespans was carried out by the HAM 310. This combined suction dredger and gravel-dumping vessel came into the field and, using her "elephant's trunk", deployed gravel at each freespan, thereby providing support and coverage of the pipelines.

Explorer II completed her workscope on 29 August, after 100 operational days during which a continuous saturation diving programme had been necessary to carry out the construction tasks.

THE 1986 CONSTRUCTION PROGRAMME

The main construction work of 1986 commenced in July with MSV Safe *Karinia* set-up over the subsea template and acting as hub for the monohull vessels and their workscopes.

Due to operational commitments, the *Installer* and the *Explorer II* were not available to carry out the work. This made it necessary for ITM Offshore to charter the DSV *Norskald* to carry out the control umbilical laying, and BUE Services chartered DSV *Arctic Seal* for the final spool piece tie-ins and control umbilical connections.

DSV Norskald

DSV *Norskald* loaded-out the nine reeled control umbilicals at Leith and mobilized to the field on 30 July. The vessel's diving capability was not used during the workscope, as all required diving services were provided by MSV Safe *Karinia* at the template and DSV *Arctic Seal* at the satellite wells.

Laying of the control umbilicals was initiated at the template by *Karinia*'s divers prerigging the pull-in wire and then passing it across to *Norskald* to be secured to the ballast wire. Once secured, the control umbilical was spooled off the storage drum, through the linear engine and tensioning monitors and thence to the sea bed. The umbilical was then tied-back to the template and *Norskald* moved away in a direction so as to achieve the specified lay route in the template vicinity and then into the lay corridor. The lay was monitored at touchdown by the Scorpio ROV, and as the vessel proceeded the "as laid" information was recorded. When *Norskald* reached a satellite well she was positioned so that the pre-terminated umbilical was laid down and the connection stabbing plate was abandoned approximately 5 m from the well.

On completion of laying the control umbilicals and carrying out the surveys, *Norskald* departed the Balmoral Field for demobilization.

DSV Arctic Seal

The largest subsea construction workscope, not directly connected with the template, was carried out by BUE Services onboard DSV *Arctic Seal*.

Load-out and mobilization of the 12 flexible spool pieces and installation equipment took place on 3 and 4 August at Aberdeen.

The spool pieces (three 2 in. × 25 m, three 4 in. × 25 m, five 6 in. × 25 m and one 6 in. × 40 m) were lifted over the storage area using a mobile crane with an under-hook height of 52 m and were then laid out on the deck to facilitate easy access for overboarding.

In the field, the schedule of work locations essentially followed that of the lay order of *Norskald*. This meant that the diving programme was continuous, with minimum time lost transitting between satellite wells.

The spool piece installations followed much the same procedures as had been devised the previous year. Small changes were necessary because of the deck-plan differences. For example, the overboarding chute was located aft on the starboard side and not over the stern, as on *Explorer II*. The sea-bed conditions and variations in elevation at each satellite well required different spool piece configurations in order to achieve the tie-in of the pipelines. At one well it was necessary to install a 25 m 6 in. water-injection spool piece with a complete loop. The elevation of the target flange ensured that there was no contact at the crossover point. Flange connections were made up in the same way with the added feature that with the larger SDC of the *Arctic Seal*'s diving system, two divers were able to lock-out and thereby greatly increase the efficiency of installations.

On completion of each tie-in, the divers installed the control umbilical, previously laid by DSV *Norskald*.

The divers' first task was to pull the umbilical stabbing plate, ballast wire (including weak-link device) and electrical cable up to the satellite well. In some cases this required the rigging of a downline through a snatch block on the guide base to pull the assembly up to the well. The hold-back wire was then clamped to the guide base and the hydraulic connection plate was lifted up ready for stabbing. Once stabbed, the plates were torqued down to ensure that all lines were connected and sealed. The electrical connection was then made. This required the setting up and dewatering of a small habitat in which the diver joined the Kintek connectors before securing the cable to the tree.

The spool piece installations and control umbilical connections at each well were followed by a series of tree function tests and the 6 h hydrostatic leak checks required on each pipeline. The tests and checks were carried out from MSV *Karinia* located over the template. The divers and ROV onboard the *Arctic Seal* carried out the statutory leak surveys at flanges and assisted with the tree function tests as a part of the completion programme for each satellite well.

On completion of the work at the satellite wells, *Arctic Seal* performed a number of remedial tasks in the field. These included the location and marking of two rig anchors and the recovery of one 30 000 lb (13.6 tonne) anchor. Her final task before departing the field was to recover a pin pile from the template vicinity, which had been installed during the 1985 programme but had become exposed above the sea bed. Divers secured the crane wire to the pile and then were recovered to the system. This was followed by the crane exerting an upward pull of 80 tonnes for approximately 25 min before suction was broken and the pile was hoisted to the deck.

The *Arctic Seal* completed BUE Services workscope in the Balmoral Field on 21 August after 16 days in the field, and returned to Aberdeen to demobilize and complete the saturation decompression.

CONCLUSION

It can be seen that monohull vessels have a significant part to play in subsea construction, both in existing offshore fields and in future developments. With the move towards subsea production systems, it will be vital to have vessels that can perform well during installation programmes and still keep costs to a minimum – an ever-increasing concern in the oil industry today.

PART II

7

Protection of subsea installations

P. H. Towers-Perkins, Kongsberg Subsea Developments Ltd

INTRODUCTION

Subsea Production Systems generally incorporate a template or satellite structure which has the following primary functions:

- location and verticality of wells;
- support of subsea components and maintenance equipment;
- limitation of deflections between components of the system;
- protection of subsea equipment.

The first three of these requirements are normally driven by system parameters, such as number of wells, method of maintenance or method of flowline pull-in and connection. Protection of the system, however, depends largely on the local environment and can have far-reaching consequences on the layout, installation and maintenance of the subsea system.

In particular, if a subsea system is installed in a fishing zone, it must be decided at an early stage which of the following solutions is to be chosen:

- a structure which will protect equipment by snagging any trawl gear on an outer bumper frame, thus preventing snagging or

Advances in Underwater Technology, Ocean Science and Offshore Engineering, Volume II – Underwater Construction: Development and Potential

snarling of equipment in the template: the trawl gear is, of course, usually destroyed;
- a structure which will deflect trawl boards and nets without damage to either subsea equipment or fishing gear.

This chapter discusses options for alternative forms of protection, and describes and compares various structures already installed or taken to detail design as examples of the different philosophies.

PROTECTION REQUIREMENTS

General

It is general practice to protect equipment on a template or satellite to some degree, in order to limit damage from typically, the following sources:

- fishing gear;
- dropped objects;
- anchor snags.

Because the wells on a subsea production system will be protected by subsea safety values, it is unlikely that damage from these causes could result in hydrocarbon spillage.

Damage could, however, result in the following:

- inability to re-enter or kill the well;
- intervention required to replace damaged equipment;
- additional inspection requirements and removal of tangled nets.

It is interesting to note that the NPD Regulations now require that all subsea installations in the Norwegian Sector of the North Sea be designed so that fishing gear will not be harmed. This requirement may not apply to fishing exclusion zones, which could be requested on the grounds of:

- low fishing activity in the area;
- proximity to permanent platforms.

No such requirements yet exist in the UK sector.

Protection against fishing gear

There are various trawling techniques commonly used in the North Sea, and among these are:

- white-fish trawling;
- industrial trawling;
- beam trawling.

Since the white-fish trawl arrangement can apply the largest snag load to a subsea structure, only this form of trawl will be described. Trawlers also deploy anchors which can cause similar loadings to trawl gear snag loads.

White-fish trawling (Fig. 1)
A white-fish trawl consists of a net with a mouth up to 25 m wide and 9 m high. The net is connected to the two towing warps by a ground line with bobbins and a headline with floats. The warps are held apart by trawl boards which can weigh up to 2 tonnes. Although the larger boards will do most impact damage, it is often the smaller, Polyvalent

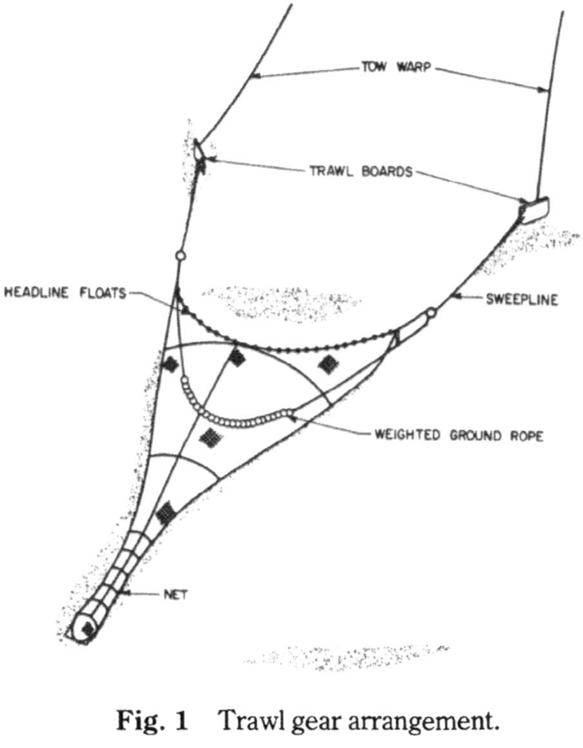

Fig. 1 Trawl gear arrangement.

trawl boards which are most likely to snag in a structure. Breaking strain for the towing warps will be up to 50 tonnes, and a double snag could therefore apply a load of 100 tonnes to a structure. The exact type of trawl deployed will vary with location, type of fish, nationality of trawler, etc.

In order to protect a subsea system, the designer has two basic options:

- to design a structure that will snag or capture nets and boards – breaking the trawl wire in so doing and destroying the trawl gear;
- to design a structure which will deflect trawl gear over itself without damage either to the trawl gear or to the structure.

The choice of protection method has significant consequences on the design and operation of the structures, and will be considered in more detail later.

Whichever type of structure is chosen, members will still have to be designed to absorb the impact energy of a moving trawl board, which can be of the order of 20 kJ. For deflection structures, which necessarily incorporate a close-fitting roofing system, and possibly hinge and lockdown mechanisms, the consequences of plastic deformation can be serious.

Dropped objects

Typical energies of possible dropped objects are shown in Table I. It can be seen that protection against a falling BOP stack or even a Christmas tree is impractical, and this possibility must be minimized

Table I
Impact energies of dropped objects

Object	*Mass (tonnes)*	*Energy (kJ)*
BOP stack	200	100 000
Christmas tree	30	10 000
Single-joint drill collar	2.5	1 000
Single-joint drill pipe	0.27	70
Drill tools	0.1	10
Small tools	0.01	1

by proper handling procedures. A drill collar energy, on the other hand, can be absorbed by extensive, but practical local protection, or even a roof, if for example a complete manifold needs to be protected. Protection against small objects can increase costs and decrease access to such an extent that the cost-effectiveness is debatable.

Protection against anchor snagging

Anchors from work vessels can impart extremely large loads, typically of the order of 300 tonnes to structure and flowlines, and it is not usual to consider such occurrences in design. A more rational approach is to reduce the risk of snagging by controlling the position and movements of such vessels, testing anchors after positioning and continuous monitoring of anchor cable tension. Small vessels are not so easily controlled, however, and it is usual to assume that an anchor from a supply vessel or fishing boat can snag certain members of the template, and possibly the flowline and umbilicals. Typical loads vary from 50 to 100 tonnes, and will be discussed in more detail later.

TYPES OF PROTECTION

General

In this section, protection will only be considered for equipment placed on or above the sea bed. Whilst it is recognized that enclosing equipment in a caisson below mudline has considerable advantages over more-conventional systems, there is not yet enough field experience in the North Sea to justify comparisons.

Any protection system will usually include both local and global protection. These are discussed below.

Local protection

Local protection on subsea production systems is used to give specified components exceptionally high protection. Because of the cost and weight involved, it is common for only master and swab valves on Christmas trees to be considered for protection against, for instance, a dropped drill collar, since loss or damage of these components may prohibit vertical well re-entry or well-kill facility.

Local protection roofs can be designed either to absorb the energy from the falling object or to deflect it elsewhere. Although the latter method can protect against greater energies, this must be balanced against the knowledge that the falling object will be deflected possibly to do further damage. All local protection has a detrimental effect on the weight, ease of running and retrieval and accessability of equipment.

Energies will in practice vary considerably for an object such as a drill collar, depending on the angle at which it falls. There appears to be insufficient data available at present to define trajectories with confidence, but protection consisting of a steel-plate sandwich of the order of 0.5 m thick would be required.

Concepts for local protection are illustrated in Fig. 2.

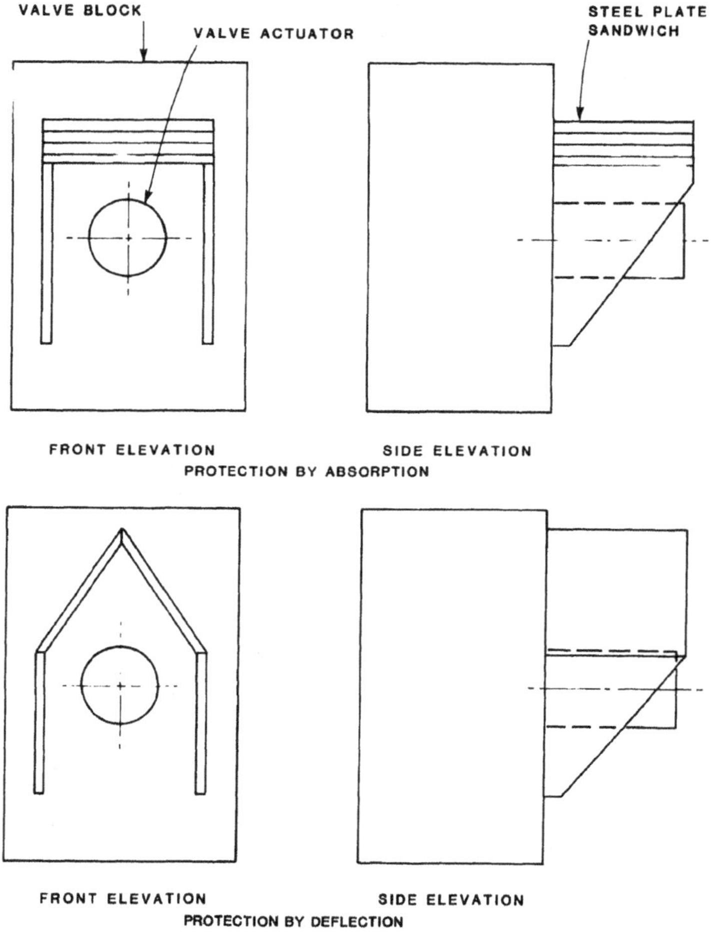

Fig. 2 Local protection concepts.

Global protection

General

Global protection may be used to protect any subsea system, and can be broadly split into two main categories:

- snagging structures;
- deflection structures.

These two types of structure are discussed below. Although only steel and concrete structures are considered, studies with composite materials have indicated that considerable material and installation cost saving can be achieved by their use.

Snagging structures

Snagging structures are usually deployed where either the area has been declared a fishing exclusion zone or there has traditionally been very little fishing. A typical snagging structure is shown in Fig. 3. The following points should be noted:

- the sloping V-bracing ensures that, once a net is snagged, it will not ride up the legs and fall into the template;
- the template does not require a roofing system to deflect trawl gear;
- the amount of steel required for the bumper frame represents about 30% of the total weight of structural steel;

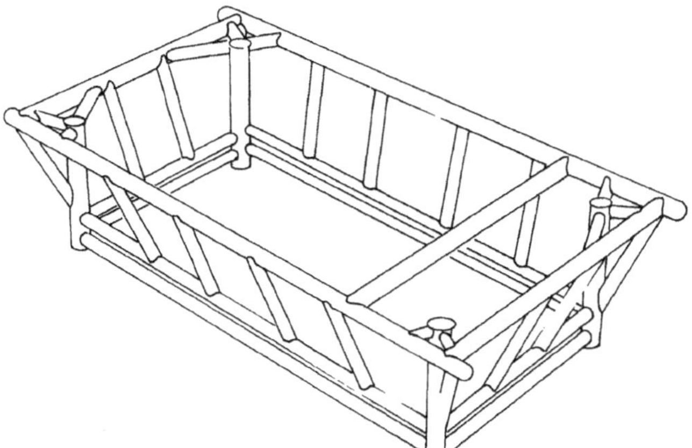

Fig. 3 Protection by snagging.

- some local protection is necessary to protect manifold valves and Christmas trees from falling objects.

Because snagging structures are often chosen for areas where little or no fishing will take place, it is assumed that flowlines and umbilicals will not be snagged by anchors within a distance of several hundred metres of the structure. In recognition of this, the design load transferred to the structure by a flowline snag is often taken to be as low as 50 tonnes. This is considerably less than the figure usually taken for deflection structures, where a flowline snag may occur within a few metres of the structure, and values of 200 tonnes are common.

It should also be noted that the upper bumper frame need not be continuous if the designer wishes to have openings for the following features:

- connection of flowlines and umbilicals;
- pig launcher running and retrieval.

However, the designer would have to accept a slight increase in probability of a trawl board or anchor entering the gap and damaging template equipment if such gaps were incorporated as permanent features.

Deflection structures

Deflection structures are deployed in areas of high fishing activity. A typical structure is shown in Fig. 4. The following points should be noted:

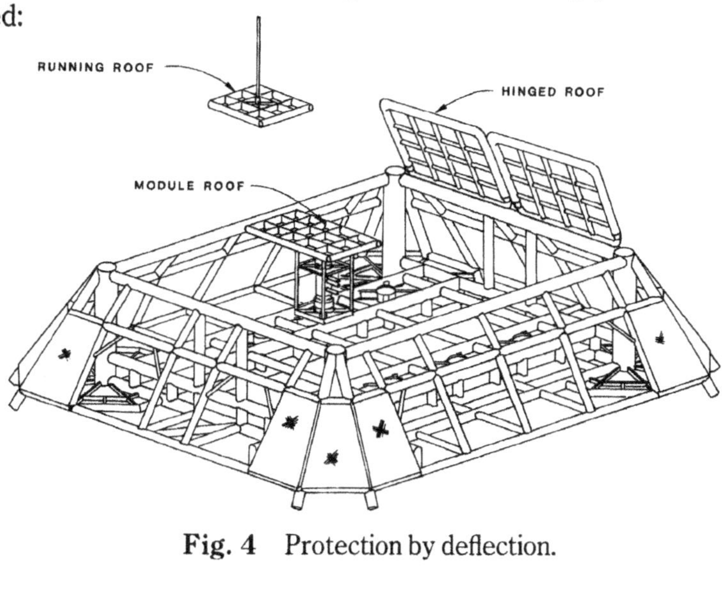

Fig. 4 Protection by deflection.

- the raking tubulars ensure that trawl gear and anchor wires are transferred over the structure without damaging either the trawl gear or the structure;
- the structure requires a complete roofing system to ensure nets do not become entangled in the template equipment: these roofs do not usually protect items below them from dropped objects;
- if the structure is fabricated from tubulars, it cannot guarantee deflection of an anchor: the main deflection members must therefore be designed to withstand an anchor snagging load;
- overtrawlability is required not only during hydrocarbon production but also immediately following structure installation.

In order to deflect trawl gear, a great deal of attention must be paid to details such as padeyes, brackets and hinges. The upper bumper frame must be continuous, and if openings in it are required for layaway of flowlines and umbilicals, provision must be made to close the gap during template operation.

Model testing is usually required to confirm the overtrawlability of the structure. This can be undertaken in standard test tanks, and entails towing a scale trawl across the structure at various speeds, angles and offsets. Due to the large variety of trawl gear and trawlers, specialist help is usually essential in this area. Tests usually reveal areas where minor modification to structural details are required.

Because a complete roofing system is required, it is necessary to remove or open a roof before any intervention can commence. The usual roof arrangements are:

- hinged roofs;
- retrievable roofs;
- integrated module roofs which form the template roof.

Hinged roofs require a lock-down system, a hinging system and an opening mechanism. The simplest are diver-operated, but these could be adapted for ROV or drillstring opening if required. The principal advantages are:

- the possibility of dropping the roof on the template equipment does not exist;
- no additional running trips are needed to open or close roofs;
- they can be larger than a roof which is intended to be run through a 6 × 6 m moonpool.

Hinged roofs are usually installed with the template, which complicates installation procedures, and recovery of a damaged roof can be complex. Retrievable roofs overcome both these problems, although they are usually installed with the template unless drilling commences as soon as installation is complete. They do, however, require additional trips to run and retrieve them each time intervention occurs. Trips can be minimized by combining the roof running tool and the guidewire establishment tool.

If the individual module roofs are designed to form the overall template roof, no additional trips are necessary when a module is retrieved, but this can make the module cumbersome to handle. Furthermore, because a complete roofing system must be in place at all times, dummy modules would need to be installed if, for instance, it was decided not to use a wellslot on the template.

Any side of a roof which forms part of the outer bumper frame will usually be designed for an anchor snagging load of 50–100 tonnes. In order to achieve this, a locking mechanism must be incorporated which could consist of:

- a bolt-type locking mechanism;
- an interference-type lock in which the roof must be lifted vertically at the centre in order to be retrieved: eccentric loads simply lift the roof against a stop.

Particular attention must be paid to the raking corner members, which form the main means of deflection. Specifically, a method of ensuring that they always penetrate the mudline must be devised. Simple stabs or extensions will suffice in soft soils, but mechanical systems may be required in stiff conditions. An accurate sea bed survey is usually required before installation, to confirm that the assumed design parameters will ensure penetration.

The piling system must also be designed to ensure that the pile does not form a snag feature if the pile is fully driven. This is achieved by:

- driving the pile below the top of leg level using an underwater hammer and follower;
- . utilizing pile sleeves set within the structural protection.

If the pile refuses and stick-up results, it can easily be removed by divers in reasonable water depths. For diverless installation, however, more-complex contingency procedures have to be devised to remove the stick-up before swaging or grouting.

SATELLITE WELLS

General

For protection of satellite wells, deflection structures are usually preferred, and both concrete and steel structures have been installed. The cost of such protection is relatively high since, without protection, the well simply requires a temporary guidebase and a permanent guidebase, which are both installed by the drilling vessel while a steel protection structure, which may weigh between 100 and 150 tonnes, must be installed by a semi-submersible crane vessel and attached to the sea bed by driven piles.

A possible alternative to this method would be an integrated protection system, in which the Christmas tree cage was itself designed to deflect trawl gear. Such systems have been designed, and appear to have definite advantages. However, since the protection is attached directly to the Christmas tree assembly, it must be moonpool-retrievable, which results in a compact structure with low energy-absorption characteristics. It also reduces access to the Christmas tree, which in turn increases maintenance costs. Table II compares the capital and maintenance costs for a piled and an integral protection structure. Figures 5–7 illustrate typical satellite well protection concepts.

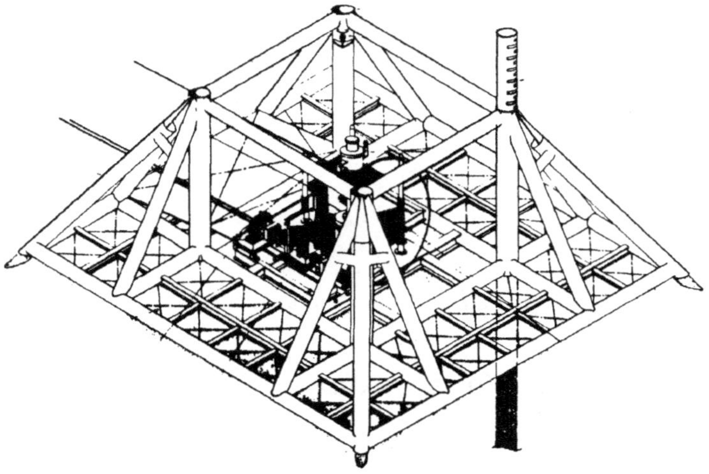

Fig. 5 Piled tubular structure.

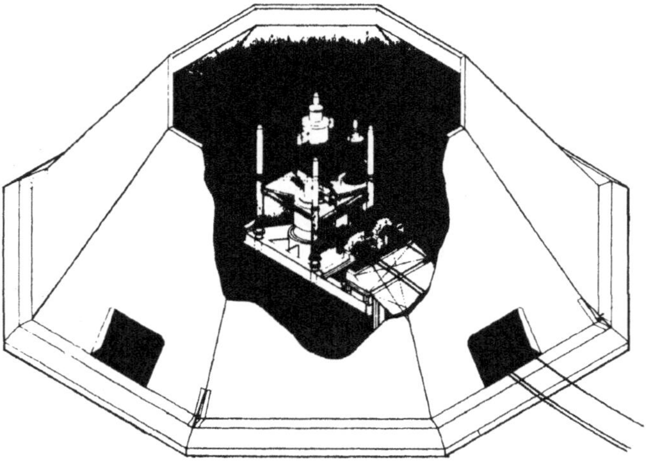

Fig. 6 Concrete gravity structure.

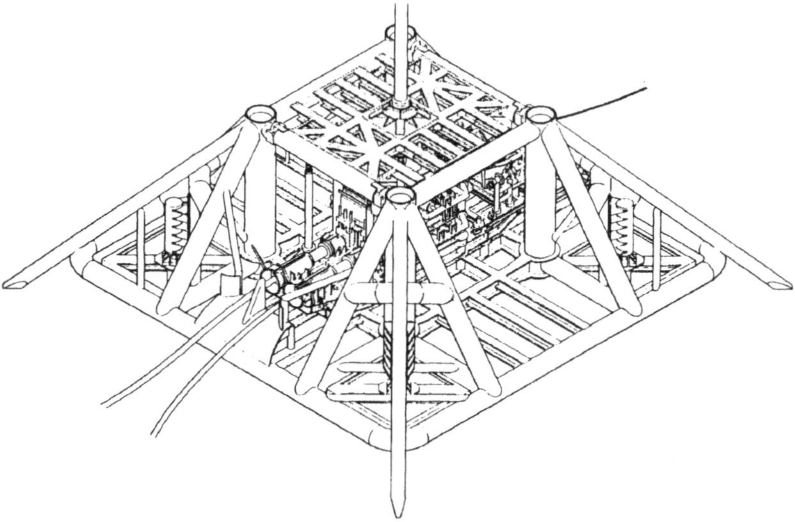

Fig. 7 Integrated protection.

Gullfaks "A" protection structures

Figure 8 shows the Gullfaks "A" subsea protection structures, which were installed in 1985 for Statoil by Inocean using the semi-submersible Karinia with a 350 tonne crane mounted.

The structures have the following features.

Table II
Comparison of satellite protection structures costs

Structure	*Design fabrication and installation (million Nkr)*	*Lifetime IMR additional costs (million Nkr)*
Integrated structure	6	24
Four-pile deflection structure	35	9
Concrete deflection structure	20	8

- The structures weigh about 160 tonnes.
- The structures are fully overtrawlable, and use a system of braces raking at 55° to the horizontal to deflect trawl gear up and over the structure. The roof is grated to guarantee that deflected objects will not fall into the interior. The grating also give a low level of dropped-object protection. Since no snagging occurs, no damage is sustained by the nets, boards or wires.
- In order to undertake diverless pull-in of the flowline and umbilicals, the upper perimeter member is not continuous. Two sides are removed with the roof.
- The roof is retrieved before well intervention by using the guidewire establishment tool.
- The roof resists anchor engaging loads by using a "jamming" system in which a load on the outside tubulars will partially lift the roof and cause it to jam, while a central vertical force will lift the roof cleanly.
- The structure is fastened to the sea bed by four swaged piles.
- Before piling, the structure is levelled to within 0.3° by a hydraulic levelling system, operated from the surface.

TEMPLATES

General

Several major subsea production template structures have now been installed in the North Sea. Some of these, together with some yet to be installed, are listed in Table III.

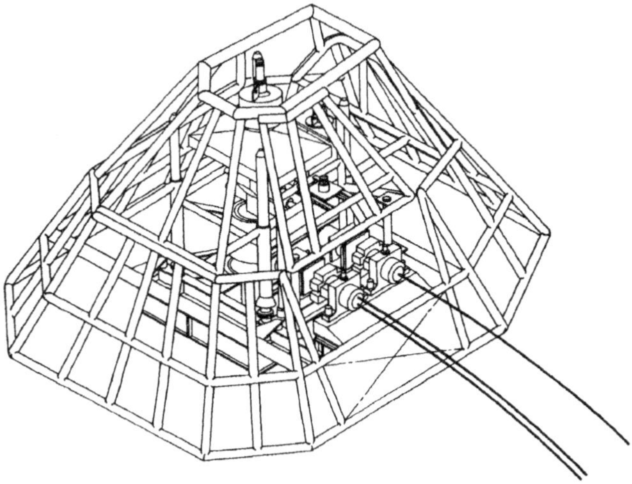

Fig. 8 Gullfaks "A" subsea protection structure.

Table III

Typical North Sea template structures

Structure	*No. of wells*	*Weight (tonnes)*	*Type of structure*
Tartan Highlander	12	1200	Snagging
Balmoral	14	380	Snagging
Shell UMC	9	2300	Snagging
Scapa	12	920	Snagging
North-East Frigg	6	240	Snagging
Frigg Additional Subsea	6	550	Snag-free
East Frigg	4	700	Snagging

In order to assess the different types of protection, the following three structures will be described and compared:

- North-East Frigg;
- Frigg Additional Subsea template;
- East Frigg template.

North-East Frigg template (NEF)

The North-East Frigg template (Fig. 9) was installed by Elf-Aquitaine Norge and uses six well slots to produce gas which is transported to the Frigg complex. The structure has plan dimensions of 21.0 × 14.5 m and a height of 10.0 m.

The template layout is simple, and is designed for good ROV

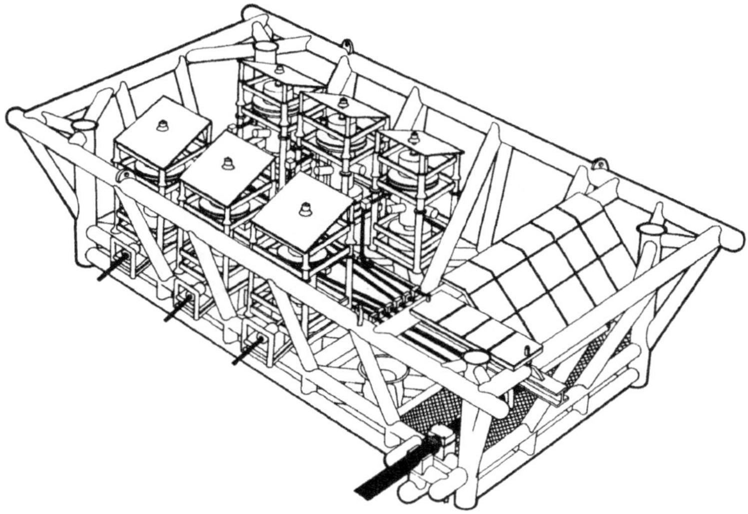

Fig. 9 North-East Frigg template structure.

access and maintenance by divers. All equipment is protected by a continuous bumper frame at the top level of the template, which runs outside the four piles to guarantee that anchor wire and trawl gear will be snagged. Vertical access is not impaired by the bumper frame, although access between the Christmas trees is reduced by the presence of the raking support members.

Christmas trees have local protection consisting of a steel plate permanently attached to the tree cap, which gives protection against small dropped objects only. The manifold area is also protected by a similar steel-plate roof.

Frigg Additional Subsea template (FAS)

The design of the FAS template (Fig. 10) was based closely on the NEF template with the additional requirement of overtrawlability. Well and manifold layouts are similar, but overtrawlability requirements lead to a structure with plan dimensions of 37 × 25 m and a height of 9 m. Overtrawlability is guaranteed at the corners by "claws" which remain buried in the sea bed even if one corner is lifted during levelling.

East Frigg template

The East Frigg template is shown in Fig. 11. It is a snagging structure, similar in form to NEF, but because of the diverless

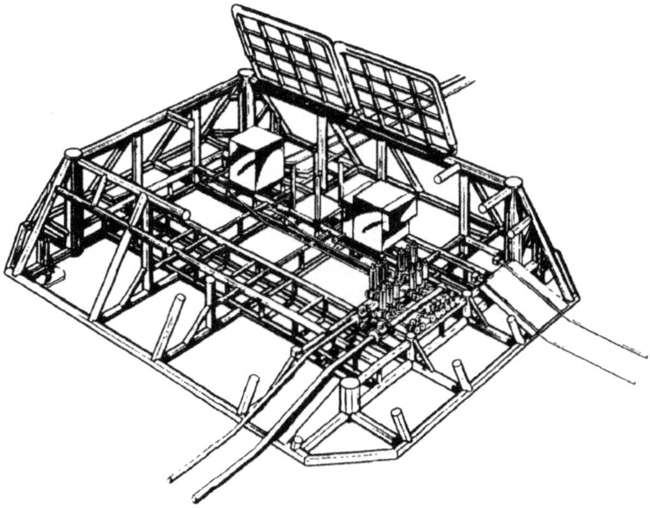

Fig. 10 Frigg Additional Subsea template structure.

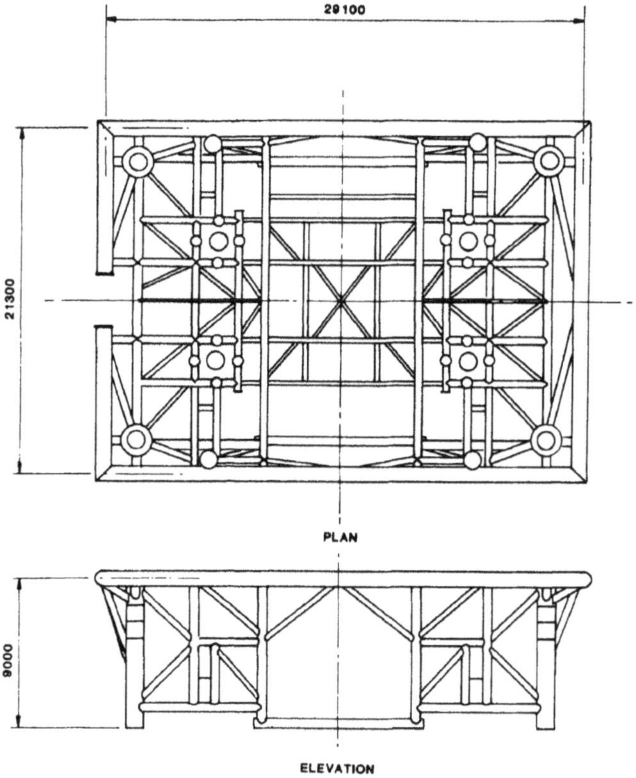

Fig. 11 East Frigg template structure.

installation and maintenance philosophy employed in the design, the plan dimensions are considerably larger at 29 × 21 m than would be expected for a diver-intensive maintenance philosophy. The structure is 9 m high and contains a break in the upper perimeter member to allow pull-in of the umbilical.

The protection frame is designed to take an anchor or trawl load of 50 tonnes and a flowline snagging load of 100 tonnes. The Christmas trees and manifold are protected by local protection with energy absorption of 50 kJ. The total weight of the template is 700 tonnes.

CONCLUSIONS

From the above descriptions it can be seen that a considerably larger and heavier structure is necessary if deflection is required, with consequent increases in material and installation costs.

While it is difficult to assess these cost increases accurately, an increase of the order of 60–80% can be anticipated in weight, plus more complex structural details and more manufactured items such as hinges, locking mechanisms and opening mechanisms.

The installation of a deflection structure requires a considerably larger vessel than the installation of the equivalent snagging structure, and often requires additional diver intervention to remove devices, to lock-down roofs during installation, and other tasks.

Many major maintenance tasks are also complicated by the presence of the roofing system. It is not, therefore, thought likely that deflection structures will become common place in the UK Sector of the North Sea unless government regulations require it, and a snagging structure with dropped object protection for selected items will probably continue as the basis for most template structures. In the Norwegian Sector of the North Sea, however, some of the structures being designed at present for installation in the deep Norwegian trench are overtrawlable.

For satellite wells a snagging structure is not normally considered, and the choice is between a deflection structure, such as Gullfaks "A" or Oseberg, or designing the Christmas tree and conductor system to withstand the stresses induced by an anchor or trawl-gear load. Again, the cost of a separate piled protection structure is considerable, but a deflection structure integrated with the Christmas tree may prove an attractive option. Such a protection system could be post-installed if it were found that the satellite well was prone to anchor or trawl snagging.

Protection against dropped objects is unlikely to increase beyond the present level of a dropped drill collar on critical areas such as master valve actuators and manifolds.

BIBLIOGRAPHY

1. S. Mathiassen, *Requirements of Protection Systems and Principles of Solutions.*
2. T. Halvorsen, *"Gullfaks 'A' diverless subsea production system". Subsea '85 International Conference.*

8

Quality assurance for subsea installations

P. R. Batten, Advanced Production Technology Ltd

INTRODUCTION

The sea bed is not the most accessible of places for dealing with breakdowns or the retrieval and replacement of defective equipment, nor does it permit frequent and easy maintenance operations.

Subsea installations are therefore designed for long trouble-free periods of operation of many years' duration. The necessity for quality assurance when purchasing equipment is very obvious.

The purpose of this chapter is to give an insight into the setting up and operation of quality assurance and control activities relevant to the provision of equipment for subsea installations. Additionally, some examples are given based on experiences during the last 3 years, of incidents which demonstrate the need for precise communication of quality requirements and the value of dedicated surveillance during manufacture and testing.

Quality assurance and quality control are not the same, and an appreciation of the difference is necessary. In the past there has been a considerable degree of misunderstanding, and it was proclaimed by some sceptics that both of these were merely fancy new names for inspection – and an excuse for delaying progress of work by introducing requirements for vast quanitities of documentation and

Advances in Underwater Technology, Ocean Science and Offshore Engineering, Volume II – Underwater Construction: Development and Potential

certification to satisfy the whims of customers and regulating authorities.

No doubt the objectives of a quality system as being the identification and specification of relevant "good engineering practice" with subsequent confirmation by monitoring, recording and certification are better understood throughout industry today than several years ago, and also the difference of meaning between "assurance" and "control", but never-the-less it might be prudent to commence by defining these terms again, and by briefly summarizing quality assurance philosophies and the formal quality system which results in the establishment of a quality plan setting out the degree of control which must be exercised.

QUALITY ASSURANCE PHILOSOPHY AND QUALITY SYSTEMS

Quality assurance is the systematic performance of all planned activities and functions necessary to provide confidence that an item or facility will be satisfactory in service.

Quality control consists of those actions which provide a means to control and measure the characteristics of an item or facility to specified requirements.

Most companies now profess the philosophy of operating a *quality system* in accordance with BS 5750 or similar standard, to whatever degree is suitable for their activities and product. This philosophy will recognize that the assurance of quality must be fundamental for all work carried out within the organization, and must be practised by all personnel in their daily activities. The existence of a QA/QC department within the organization does not diminish the individual quality related responsibilities of other personnel.

The philosophy will also recognize that quality is enhanced by working in a disciplined manner to formal procedures designed to eliminate the occurrence of deficiencies, and that the evidence of quality, before the self-evidence of satisfactory performance in service, is provided by records generated during any activity undertaken towards provision of the product.

To establish the required disciplines within its quality system a company will develop a *quality programme* consisting of a documented set of activities, resources and events serving to implement that system. This will provide the recorded evidence of quality by means of:

- written procedures which detail the purpose and scope of an activity, and specify how it is to be properly carried out;
- documentation which records the results of tests or activities;
- certification which confirms that an item, test or service conforms with specification.

A company's general quality policies, procedures and practices are usually summarized in a *quality manual*, which serves as a guide for its own personnel and as an introductory document from which a potential customer can make a preliminary assessment or use as the basis for a formal quality audit.

The foregoing is a brief outline of the generally accepted structure of a quality system, and assuming that a supplier has been judged to have an ostensibly organized and effective system, can it also be assumed that the responsibility for supplying equipment which is fit for purpose rests entirely with him through his Quality Assurance Department?

Contractually, the supplier is responsible for supplying the equipment to a quality level no less than that which has been specified in the order, and there will be clauses stating that no action or decision made by a customer's representative during surveillance or witnessing functions shall relieve the supplier from any part of this responsibility.

This does not mean that the purchaser's own personnel are exempt from any responsibility towards the achievement of required quality in the equipment which their organization will eventually be using. This achievement results from combined effort by all parties, and a considerable personal responsibility for some individuals.

In practice the quality process is initiated by the engineers who are identifying and specifying the operating conditions and requirements. Thereafter it becomes the collective responsibility of everyone concerned with specifying, designing, purchasing, manufacturing and accepting the equipment, and good communication and co-operation are essential.

THE NEED FOR SPECIFICATION OF QUALITY REQUIREMENTS

The problems of repair or replacement of equipment on the sea bed, and the consequent cost in terms of lost production, workover and rig time, is very much higher than on a platform and can be expressed as *very* many times the direct cost of the failed item. Although the direct cost of failed equipment might be recovered from the supplier, the much larger consequent cost would be borne by the customer.

The need for thoroughness in identifying critical areas, the clear communication of requirements to the supplier, and the introduction of meaningful quality and inspection specifications for subsea equipment becomes very obvious.

The application of such standards in the form of a contractual *quality plan* will also serve to commit the supplier to a specified quality level for a stated price, and no unforeseen costs should be incurred at a later date because of misunderstandings concerning requirements. Additionally, it becomes possible to inspect and monitor progress during all manufacturing and test stages, since criteria and standards have been defined.

QUALITY REQUIREMENTS

The identification and specification of quality requirements are the first steps towards effective quality control. Any number of engineers within an organization might make an input to these activities, and a disciplined approach to the task is essential in order to arrive at an integrated result expressed in unambiguous and recognized quality terms.

For this reason it is advisable for the customer's organization to create very comprehensive procedures and guidance notes for reference at this initial stage. These should provide a checklist of virtually all activities which may require attention, and sufficient description of the scope of such activities to assist the engineers in deciding exactly what is necessary to suit the criticality of the equipment or item in question. The importance of giving this task the care and attention which it merits cannot be overstressed, and the engineer must assess whether it is necessary to invoke all or only relevant parts of the check list or whether, in some cases, additional requirements must be added.

The quality requirements must be carefully considered for their value and impact on the reliability of the finished product, and should be thoroughly reviewed in total before being compiled as a specific section of the enquiry document to be sent to prospective bidders. Typically, a quality requirements section will be constructed as follows.

1. Introduction

A brief summary of the relevance of the attainment of quality, and hence reliability, to the equipment which is to be supplied. Also the impact of statutory certification if this applies.

2. Definitions

Key words and reference to relevant standards, e.g. BS 4778 "Glossary of terms used in quality assurance".

3. Draft quality plan

Bidder is required to submit a draft quality plan describing the resources and activities which will be employed to achieve quality objectives for the specified scope of work. This should include descriptions of the quality control system and procedures proposed for:

- design control;
- change control;
- planning control;
- material control;
- procurement of materials and services;
- control of free issue material;
- manufacturing control;
- control of sub-contracted items and services;
- special processes (e.g. welding, heat treatment, protective coatings);
- inspection functions;
- proving, factory acceptance and interfacing tests;
- non-conformances and rectifications;
- control of inspection equipment, gauges, recorders, etc.;
- weight control;
- preservation, packaging and despatch;
- commissioning procedures;
- operating and maintenance manuals;
- documentation, records and certification.

The draft quality plan should also include a manufacturing schedule identifying key points where quality-related activities, inspections and tests take place, and identifying proposed hold and witness points.

A list of required documentation, records and certification should be included in the quality requirements section. Typically these may include:

- design specifications, calculations and justifications;
- as-built drawings;
- welding procedure specifications and qualification reports;
- welder qualification records;
- heat treatment or other special process procedures;
- material certificates;
- NDT (non-destructive testing) procedures;
- NDT operator classification records;
- inspection and test procedures;
- inspection and test records and certificates;
- heat treatment charts;
- approved concession notes, change notes and repair weld procedures and reports.

The quality requirements section must also emphasize any other aspects which will become binding in the event of an order or contract being placed, e.g.:

- Vendor's responsibilities – to ensure that all activities and functions necessary to achieve the specified quality are properly administered and carried out. Customer's representatives may carry out monitoring, surveillance, inspection and witnessing, expediting, or quality audit activities during the course of the order or contract, but none of these shall detract from the vendor's quality responsibilities.
- The vendor shall nominate a senior person to be responsible to the customer for the satisfactory resolution of all quality-related matters.
- The vendor is responsible for quality-related activities carried out by any of his sub-contractors.
- In any case of doubt concerning the meaning or intent of specifications, drawings or contract documents, the vendor shall direct all enquiries to the customer for clarification.
- No deviations from specification shall be undertaken without prior discussion with and agreement by customer.
- The customer shall nominate a representative or representatives with a specified scope of work, who shall be accepted by the vendor and allowed all reasonable access and assistance at the vendors' premises and at the premises of his sub-contractors to allow the performance of designated duties.
- All documentation and certification generated in accordance with the quality plan shall be checked by the vendor's relevant personnel and collated into a QA Dossier at the time that it is produced or received, thus forming a continuously current

document available for checking at any time during manufacturing or test activities.

QUALITY PLAN AND QUALITY CONTROL

Ultimately, following discussions between the bidder and customer the draft quality plan will be refined into a mutually acceptable plan which will form part of a contract for manufacture and supply. This indicates that a genuine effort has been made to set up a comprehensive system of quality control detailed and documented in such a way that, if adhered to, the declared objectives will be achieved. Can we assume with confidence that this will be so? After all, the quality assurance documentation package to be supplied with the equipment will provide all the evidence necessary to prove that quality control has been properly carried out. This will be checked by the customer, and there is probably a clause saying that final payment will not be made until the documentation is received and approved. Excellent; the supplier now has a financial pressure to ensure that he supplies proof of quality, it is in his own interest, and if something does go wrong the customer has a stick with which to beat him! Such an attitude is naive, but unfortunately it is not uncommon.

Undoubtedly, all that has been done in documenting the quality plan does represent the disciplined application of good engineering practice, and its impact is designed to reach all those areas where particular care is required, but quite frankly it would be irresponsible to assume that the adoption of the quality plan will automatically guarantee success.

Common sense and experience will recognize that, even within the most well-regulated systems, human error and lethargy cannot be entirely ruled out. For this reason there is no more effective way of enhancing quality control than for a customer's representative to augment liaison by site visits and to demonstrate his personal commitment by "donning a boiler suit" and carrying out objective activities to check progress and adherence to the quality plan.

An experienced man will very soon assess the true situation and take appropriate action, and can, in the pursuit of his objectives, be useful to the supplier as well as the customer. More subjectively, but of considerable value, the right person can, by example, instil a sense of commitment into others.

The very real value of such liaison, monitoring and surveillance activities can possibly be illustrated by the following examples of problems encountered and resolved during the manufacture and supply of major items of subsea equipment during the period from May 1984 to May 1986.

SITE EXPERIENCES

Welding Procedures and Qualification Records

A common problem was experienced with a number of suppliers.

The difference between a Welding Procedure Specification, a Procedure Qualification Record and in some cases a Welder Performance Qualification Record is not always well understood, even by some Certifying Authorities' personnel, and inadequate documents are often submitted for approval in connection with a particular weldment.

The most usual mistake is the supposition that a Procedure Qualification Record is a Welding Procedure Specification, and the lack of realization that a Welding Procedure Specification should be specific for a required joint thickness and geometry, and provide the necessary instructions to the welder for making that joint.

Any number of Welding Procedure Specifications can be generated from and upheld by an acceptable Qualification Record, and a fabricator should not, for example, submit a full-penetration vee-butt weld procedure and qualification record for use on a fillet weld under the justification that "qualification on a butt weld covers for fillet welds using the same welding process". The qualification record covers him to write an appropriate fillet weld procedure specification within the constraints of the relevant Code or Standard, and this is what should be put forward for approval.

It has been possible to rectify this in a few cases, but in general it remains a problem throughout the engineering industry, and will only be solved by gradual and concerted insistence by customers when documenting their quality requirements for new work.

Unauthorized weld repair

During an unscheduled visit to a supplier's welding shop approximately 4 days after a previous visit it was noticed that welding was still being performed on joints between flowloops and a major component forging. The welds should reasonably have been completed shortly after the first visit, and the welder, on being questioned said that he was repairing weld defects shown up by NDT.

The contract required that weld defects should be reported and a repair procedure approved by the customer before repair.

This had not been done – not, apparently, deliberately in order to conceal the need for repair, but because the technician responsible for the control of welding was unaware of the contract requirements.

The welder, more accustomed to such mandatory requirements, had "assumed" that the technician had received clearance from the customer. The supplier's Quality Manager had thought that the Quality Engineer who produced the quality plan had ensured that everyone concerned had been advised of the repair weld restrictions, and aged about 5 years in 5 minutes when the incident was reported to him.

The matter was resolved at a meeting between the customer's representative and the supplier's design engineers. A suitable modification was proposed, checked and authorized, and the necessary work was satisfactorily completed within 10 days.

Receiving inspection of free issue items

A supplier decreed that while he would check free issue items against the advice note for quantity and description, he could not be responsible for condition of items free issued by the customer for incorporation into the equipment, and would assume they were correct to drawing.

The customer opined that the supplier should check all incoming goods, irrespective of source, but admitted that, in fact, it would be difficult to obtain drawings from the manufacturers of the free issued items, so in reality it would be impossible to make any significant dimensional or surface-finish checks.

The part numbers etched on the components were correct, however, and it was therefore a reasonable assumption that the original manufacturer would have checked them anyway.

Subsequently, long after the components had been welded into complex pipework and assembled into a three-quarter million pound subsea unit, it was discovered during customer witnessing of a TFL (through flow line) tool trial at the factory that an internal location shoulder had been machined approximately 2 in. out of position by the original manufacturer.

To the supplier's credit, he immediately and at considerable cost made up special-purpose boring equipment which allowed the components to be re-machined successfully *in situ*.

With hindsight it was admitted that free issue items of such importance to the final equipment should not be accepted without a dimensional inspection, and that in the absence of drawings it would still have been very easy to have borrowed an appropriate tool and physically tried it to ensure that it located properly into the internal profile.

Fabrication welds

Part of the base structure of a major item of subsea equipment was unbolted during pre-offshore preparation. The fitter carrying out the work reported that there was something underneath which looked like a cracked weld.

Ultrasonic investigation revealed that the "crack" was, in fact, a total lack of fusion at a depth of no more than 8 mm below the surface in what was from outward appearances a full-penetration butt weld between two 50 mm plates (see Fig. 1).

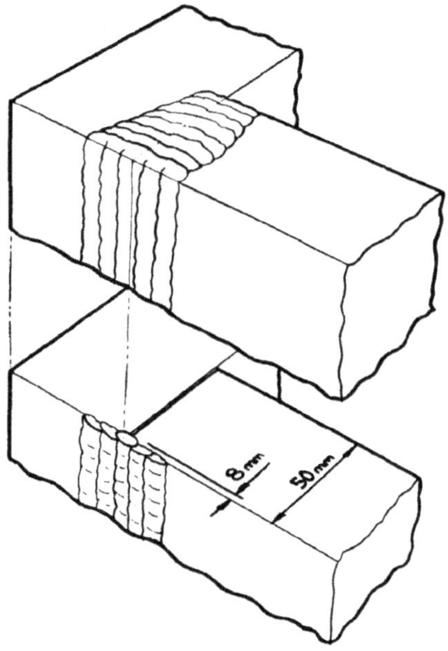

Fig. 1 Fabrication welds. Weld in 50 mm plate "cosmetically" covered on the outside to appear like a full-penetration weld.

Excavation confirmed the ultrasonic report, and a similar pattern of inadequate welding was then found in similar welds on the other corners of the base structure and on three identical units.

The welding had not been carried out in Great Britain, but the British branch of the supplying company was called in and could not obtain any information concerning the quality control, or rather lack of it, exercised by their parent company during manufacture. The investigation revealed that the plates had not been prepared in accordance with design drawings to provide a vee groove, and the welder had simply provided an artistic impression by applying false capping runs, and simulated weld runs at the ends of the joint faces. Quite evidently there had been no supervision or quality control whatsoever during manufacture.

The British subsidiary was called in to submit an *in situ* weld repair procedure and to carry out the work. Strength calculations were a requirement for the repair proposal.

Incidents such as this reinforce the need for attention to detail in the acceptance and control of joint-specific weld procedures outlined in the first example of these site experiences.

Valve seat hardfacing damage

During pre-offshore testing of a subsea gate valve persistent leaking led to it being opened for investigation.

A chipped seat was discovered, and it was then relatively simple to reconstruct a TFL operation and identify that the accumulated tolerances of the various components meant that it was possible for a considerable area of the seat or gate face to be exposed to the passage of TFL tool strings.

Surprisingly, although the TFL capacity had been a requirement of the original equipment specification, the potential liability of the seat and gate manufacturing tolerances had not been noticed.

Approximately 36 valves of this type were involved, and a crash programme of redesign and refurbishment had to be carried out.

Hydraulic couplings, design change

An order of considerable value was placed on an overseas company for a quantity of hydraulic couplings. These were to be as previously supplied.

When received it was noticed on all of them that an annular recess had a circumferential weld, and that the faces of the recess were marginally out of square with each other. The overall effect was that a split keeper ring assembly would not fit properly into the recess.

Thinking that a machining operation might have been missed the customer contacted the supplier for an explanation, and it was arranged that the recesses would be squared up and the weld crown removed. This would be done by a British subsidiary.

The work was carried out, but it was then noticed that areas of undercut were left after the weld crowns had been removed flush with the bottom diameter of the recess.

Further investigations revealed that the previously supplied couplings had not been welded, but that for this order an electron-beam weld had been introduced. This had been done to facilitate an internal valve seat machining operation. The customer had not been informed of this design change, but the supplier maintained that because a sophisticated welding process had been used there was no possibility of a weld failure.

A thorough investigation by the customer's welding engineers, backed by the opinion of the Welding Institute, revealed that the weld configuration of a flat butt weld between components of considerably different wall thicknesses was inherently bad, that the fusion zone geometry was incorrect and liable to result in incomplete root fusion, as evidenced by several sections taken from welds, and that variations in the alignment of the beam to the joint interface would further aggravate the latter problem so that random areas of lack of fusion might extend through as much as 50% of the minimum wall thickness (see Figs 2 and 3).

Briefly, it would have been difficult to have designed or carried out a more unsuitable weld for the purpose, and there was no satisfactory NDT method which could be applied.

The supplier could not give evidence of any satisfactory welding

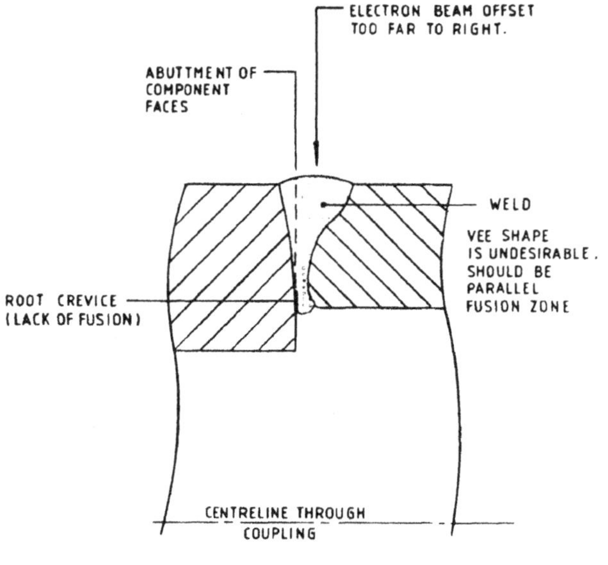

Fig. 2 Hydraulic coupling – unacceptable electron-beam weld.

Fig. 3 Photograph of the weld root of Fig 2, showing lack of fusion, resulting in crevice and potential stress raiser (× 160).

procedure qualification test and, in the face of the customer's urgent requirement for couplings, was obliged to replace the entire batch with couplings of the original non-welded design.

These were manufactured and supplied by the British subsidiary within 2 months.

CONCLUSIONS

It is a fact of life that subsea equipment needs to be fit for purpose before it is put on the sea bed, because the potential cost of replacement due to a failure or malfunction can be very high.

The assessment of the quality requirements to suit the functions of individual items of equipment has to be a dedicated activity, and the translation of these requirements into a quality plan of whatever degree of detail may be necessary, must be thought of as nothing more or less than a properly disciplined approach to sound engineering practice. The same applies to the quality control activities which are specified in the quality plan.

Documentation and certification for the final package should be limited to essentials, but it must be meaningful and complete. Documents should be checked off and signed at the time when they are relevant to design, manufacture or test, and collated as an on-going activity during the entire course of the order or contract, not gathered together as an afterthought.

Dedication to the required tasks, and co-operation between individuals and companies in pursuit of the required quality objectives, should be expected as normal from the disciplines of professional engineers and a competent workforce, but this is an ideal, and the facts of life must be faced.

The half-dozen "incidents" outlined in this chapter are fact, and are recorded together with many similar events in the daily log of one person who has assisted in establishing the quality requirements for several subsea projects and subsequently been involved with the outcome at supplier's premises and on site.

They all provide the same warning. Never, ever, assume that things will go according to plan, or that instructions have been understood all the way down the line.

It is necessary, no matter how precise the plan, to check, and check again, that it is working out as intended, and to do that you must sometimes go where the action is, and assist and observe.

9

Geotechnical influences on subsea installations

E. L. James, Fugro Ltd

INTRODUCTION

Most forms of subsea installations have, as a common feature, a requirement of physical contact directly or indirectly with the sea bed. The nature of that sea bed greatly influences the design and installation technique adopted, and consequently the economics of the project. Furthermore, incorrect assumptions about the sea bed can necessitate expensive remedial measures, since little flexibility is available to adjust the design at the time of installation.

Despite these factors, site-specific knowledge of the sea bed is often unavailable for subsea installations. Designs based on soil conditions at platform sites many kilometres away are not uncommon. This chapter describes some of the influences that the geotechnical conditions can have on the design and operation of subsea installations, and demonstrates the desirability of the acquisition of site-specific soil data.

FOUNDATION LOADS

For the purposes of classifying their basic foundation loads, subsea units can be divided into three categories:

Advances in Underwater Technology, Ocean Science and Offshore Engineering, Volume II – Underwater Construction: Development and Potential

(a) units which are set completely below sea-bed level;
(b) units which are set at or slightly below sea-bed level;
(c) units which float above the sea bed and are anchored to the sea bed.

In each case the purpose of the foundation member is to support the loads imposed by the unit and any external forces applied to the unit safely, effectively and at an acceptable deformation. The first stage in deciding on the type of foundation unit to be adopted (and hence on the soil data requirements) is to establish the magnitude of the loads for the various options under consideration.

Fully recessed sea-bed units are currently only adopted where local conditions render surface units vulnerable to damage. Typically, silos may be adopted where iceberg scouring of the sea bed can occur. Such units, by their nature, do not attract high accidental lateral loads, and silos are generally able to support their own weight and that of the unit by external-skin friction very adequately. They therefore provide their own foundation, and the only geotechnical areas for particular study are those governing their installation.

Surface units represent the vast majority of existing and planned subsea installations. The first generation consisted of a wellhead with a rigid riser and a production tree affording well control. The next stage was the development of manifold units to permit collection from a number of satellite wells and to allow transportation to adjacent topside production facilities. Current and future developments include sea-bed separation units, thus making possible long-range sea-bed transportation and the use of modular 1 atm. units serviced by submersible. Unless requiring the special protection afforded by category (a) above, these units are all normally placed on the sea bed and require support by the sea bed.

Vertical loads generated by these units and their protection cages are generally much lower than the range of loads normally associated with platforms. Vertical components of foundation loading arise from:

- self-weight of unit;
- riser forces;
- weight of well casings.

Lateral forces may be developed due to:

- bending forces arising from lateral excursions of rigid risers;
- accidental snagging by trawl boards or dragging anchors;
- flowline pull-in forces;

- environmental influences, such as currents;
- sea-bed instability.

Whereas some of the above forces can be accurately predicted, others require an element of risk assessment to establish design criteria. Of particular interest is the potential lateral force caused by a dragging anchor. The current range of construction barges carry anchor wires capable of imposing loads in excess of 600 tonnes. The design of wellhead cage foundations to withstand such a force could prove prohibitively expensive.

Guidelines are required to ensure that the foundation designer is provided with realistic forces. Codified guidelines are also required on the factors of safety to be applied in design. Stress factors frequently in use by subsea structural designers vary between 0.7 and 0.9 of yield, depending upon the nature of the load case considered, but the author is unaware of similar guidelines for the foundation design. The Department of Energy/API approach used for platforms may be inappropriate (and overconservative) when applied to unmanned sea-bed units subjected to low environmental forces. The use of a modified load factor in the NPD partial-factors approach may provide a better basis for design. In this method a material factor of 1.3 is applied to the design to allow for variations in soil parameters, and a load factor is imposed to allow for unknown aspects of the loading. Consideration should be given to the most appropriate value of the load factor, normally equated to 1.3 for platform foundation design using total-stress methods.

The loads imposed by a guyed buoyant unit on anchors will primarily be governed by the buoyant force and by lateral environment forces. Historically large factors of safety have been adopted in the design of permanent anchoring points in the sea bed.

FOUNDATION TYPES

All subsea foundations can be classified as either footings (mudmats, gravity base, bottle leg, spud can, etc.) or piles. The silo is seen as a special case of the large-diameter pile.

The true surface mudmat provides an inexpensive and effective means of supporting vertical loading, and is ideally suited for subsea applications where vertical loads predominate and where surface soils are competent. However, such a mudmat cannot mobilize significant lateral resistance. The addition of skirts or dowels below the mudmat increases this resistance by forcing a shear failure within the soil mass.

Piles enable high vertical forces to be transferred to deeper, more-competent strata. They also provide high lateral resistance. The piles used for subsea units are invariably steel-pipe piles installed by either driving or grouting into a drilled shaft. Penetration depths of between 15 and 40 m for 0.76 m diameter piles are typical.

GEOTECHNICAL INFLUENCES

Geotechnical aspects of the sea bed significantly influence the design, the process of installation and the operational behaviour of subsea units. Some of the most important influences are described in the following sections.

Regional Aspects

Most subsea units are placed sufficiently close to existing facilities for the regional geotechnical aspects to be well understood. Shallow geophysical surveys will have highlighted buried channels and other geological anomalies that are best avoided when siting subsea units. Similarly, surface phenomena such as pockmarks, iceberg scour and surface boulders can be identified by a side-scan sonar survey.

In shallow waters high bottom currents can give rise to sediment movement and redeposition as mud or sand waves. These can be detected in close-grid bathymetric surveys, particularly if periodic surveys are available from which changes can be identified. Such movement can lead to a transient partial burial of subsea units, followed by a period of scour which could leave foundations exposed.

The large-scale mobility of the sea-bed in the regions of the great river deltas such as the Mississippi gives rise to particular problems related to the volume of material transported. Information derived from shallow seismic surveys can enable areas of diapirism and sea-bed sliding to be identified.

Future subsea developments are likely to take place in the deeper waters on the continental slopes. Whilst the slopes are relatively shallow, generally about 5°, the material of the sea bed is often fine grained and of a very soft consistency. Seismic surveys of these areas have indicated frequent deep-seated rupture surfaces. Sea-bed instability stems from some triggering mechanism, and it is likely that these historical ruptures were caused by seismic activity. Examination of shallow geophysical records and the dating of shallow sediments can enable the age of these ruptures to be estimated. It is not inconceivable that the extraction of hydrocarbons without controlled water injection could trigger instability.

No foundation or well design would be able to resist such a slope failure, but a knowledge of the potential for movement is essential in seabed siting studies and in the design of well protection.

Site-specific aspects

Design methods for seabed foundations are well documented elsewhere (American Petroleum Institute, 1984). This chapter restricts discussion of design aspects to those areas of particular relevance to subsea foundations.

Mudmat design

The behaviour of a mudmat under vertical loading is governed by the stratification and the soil shear strength over a depth at least equal to the least lateral dimension of the mudmat.

The presence of a soft surface layer may lead to unacceptable penetration or tilting of the unit. A thin, firm crust overlying a soft stratum could provide adequate support under operational conditions but fail under certain extreme combinations of vertical and lateral forces or under cyclic loading.

Long-term settlements are generally not critical in the design of subsea units, provided that provision for differential movement has been made in pipeline tie-in design.

The lateral resistance of a "smooth" mudmat on a sandy sea bed will be governed by the angle of surface friction and the distribution of contact pressures. The resistance of such a mudmat may, under anchor snagging loads, be reduced to zero.

Skirt design

The inclusion of skirts and/or dowels is a standard technique for increasing the lateral resistance of mudmats by mobilizing the shear strength of the soil over the skirt penetration depth. However skirt penetration is itself governed by the soil's resistance to penetration under the available insertion thrust. The use of cone penetration tests provides data for predicting this penetration. These tests involve pushing an instrumented probe into the sea bed and recording the resistance mobilized at the tip of the probe and over its sides. By the use of empirical correction factors, acceptable predictions can be made as indicated by a recent Norwegian test (Tjelta *et al.*, 1986).

Significant variability of the upper few metres across the area of a subsea unit can lead to unacceptable tilt of the unit during emplacement unless adequate provision for differential ballasting is available.

Pile design

The design of piles under axial loading requires shear strength data for cohesive strata and *in situ* density measurements for sands, together with an assessment of the *in situ* state of stress. Such data should be available over the full depth of the piles plus a minimum of 5 m. Since only the required capacities are known in advance, it is desirable to compute pile capacities during the progress of data acquisition. Occasionally the axial load-deflection response of piles is required for the structural design of the unit.

Piles under lateral loading generally mobilize their resistance from the upper 15 diameter lengths below the mudline in North Sea soils. Where soft sea-bed soils are found this length may be increased and the lateral response of the piles may well govern the pile penetration required. Since local and general scour lead to exposure of the upper part of the piles, a softer lateral response will result. Knowledge of the likely magnitude of scour is therefore required at the design stage. This will depend as much on the nature of the sea-bed soils as on the hydrodynamic changes arising from the placement of the subsea unit.

Installation process aspects

Mudmats

The mudmats of subsea units form the basis from which the unit has to be levelled to tolerances frequently as close as 0.5° to the horizontal. Since the amount of adjustment available is restricted, some reassurance is needed on the likely inclination at set-down. This can be influenced by the slope of the sea bed, the presence of obstructions or hardspots and variations in skirt penetration resistance around the periphery.

Piles

Platform piles installed in the North Sea over the past decade have almost invariably been driven. This is due, in part, to the development of a range of powerful and very effective underwater hammers. These have helped to reduce the time needed to complete the piling of a large jacket from about 4 weeks to 4 days.

In principle, the same advantages could be gained when driving the piles of a subsea unit. The range of hammers includes a number of underwater versions that are perfectly capable of driving piles to the penetrations needed for subsea units. These hammers can be handled relatively easily from a wide range of construction vessels without the need for heavy-lift cranes and bulky power packs.

Reliable predictions can now be made on the driveability of piles from a knowledge of basic soil parameters and hammer stress-time output. Measurements of stress and acceleration made on piles during underwater driving can now be used as input to numerical solutions based on the wave equation.

The driving of subsea piles is normally the quickest and most cost-effective method for their installation. It does, however, have one major drawback; significant vibrations may be transmitted to the unit being piled. It is estimated that, by physical contact between the pile and its guide, axial accelerations of up to 50g may be imposed upon the guide. For basic templates, guide posts and wellhead cages this should not present a problem. However, the current trend is to simplify installation procedures by assembling templates and manifolds as single units onshore before shipment to the field. Care has to be taken to ensure that the sensitive valving and instrumentation on such manifold units is not damaged during pile installation. The attenuation of pile-guide vibrations to the manifold area will depend upon the structural design and the damping of the soil supporting the unpiled template. Careful consideration must be given to the design and support of all appurtanences.

If driven piles are thus found unacceptable, then consideration can be given to the installation of drilled and grouted piles. These have the advantage of using exactly the same installation technique as is generally used for the installation of well conductors. The steel-pipe pile is grouted into an oversized shaft drilled by conventional oil-well techniques. This technique is also not without its problems. During drilling, washout of surface material or of other deeper layers can occur, leading to surface cavitation and greatly increased grout take. Hole instability can lead to difficulties in setting the pile in place and in ensuring effective subsequent grouting. Since drilled and grouted piles involve relief of *in situ* stresses, the design criteria differ from those for driven piles, and consequently greater depths of penetration are required to achieve a given design capacity.

Operational aspects

Subsea units incorporating wellheads create conditions in operation which can influence foundation design. The process of drilling a well can lead to washout below the wellhead template which, at best, modifies foundation design assumptions and at worst could given rise to wellhead tilt or increased lateral flexibility. Careful planning of the methods of installing the conductor string and the first well casing, taking into consideration actual soil conditions, should eliminate this problem.

The problem of disposal of drill cuttings has finally been identified by the oil industry. Methods are now available for predicting the likely distribution of cuttings discharged at a given height over a prolonged period. Unless dumped in a designed manner, such cuttings can increase the loading on a mudmat and affect local scour regimes.

In operation, the well-casing temperature will rise, leading to expansion. Where conductors are rigidly connected to the wellhead frame (either to provide support or due to the nature of its design) the expansion will tend to lift the wellhead. This can lead to high compressive stresses in the well casings, and to high bending and shear stresses in the template. The amount of movement at the surface for a "free" well head depends upon the soil conditions around the casing as well as on factors such as temperature gradients. Soil-structure interaction solutions are available for analysing such conditions whilst taking into account the wellhead flexibility.

DATA REQUIREMENTS

This chapter has identified numerous areas where geotechnical aspects can influence a subsea development in addition to basic foundation design considerations. The problems associated with these can be minimized by the acquisition of suitable regional and site-specific data.

The first requirement of any such development is a comprehensive geophysical survey. This should provide shallow seismic profiles at a fairly close grid over any areas where units are likely to be placed, and at a wider spacing over the general area. At the same time bathymetric and side-scan sonar surveys should be conducted. The results will assist in site selection and can provide data for design if the survey can be linked to existing borehole data. In some circumstances, where the nature of the proposed facilities allows, the geophysical data may suffice for design.

Generally this would not be the case and a programme of seabed testing and sampling should be planned. A range of options are available, some of which are shown in Figs 1 and 2. Most of those shown on Fig. 1 can be operated from the type of vessel used for

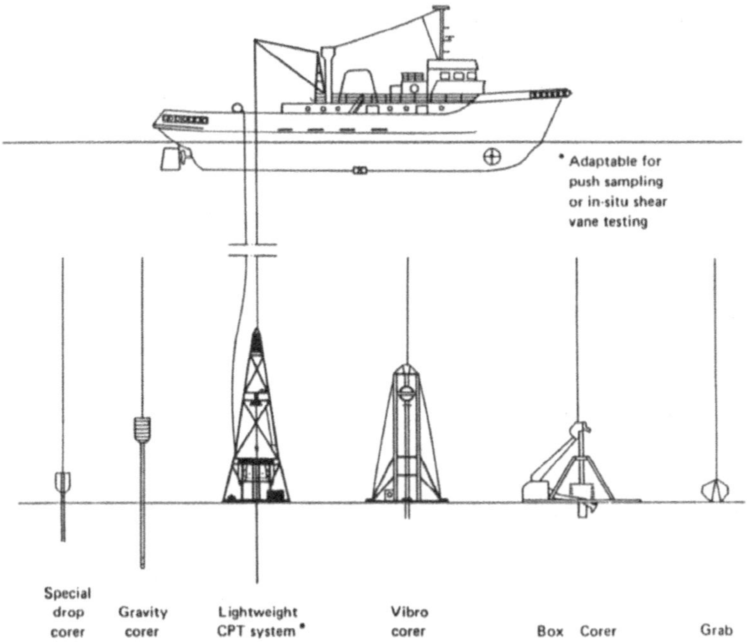

Fig. 1 Sea-bed exploration techniques operated from survey vessels.

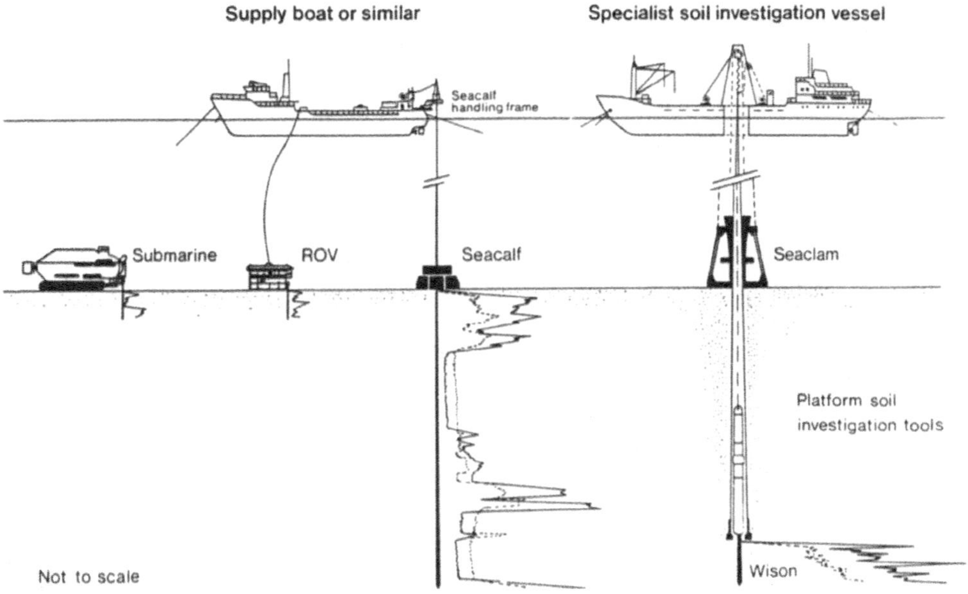

Fig. 2 Sea-bed survey techniques operated from site-investigation drilling vessels.

geophysical surveys. They provide samples and *in situ* data to a depth of up to 10 m, which may suffice for some mudmat-supported units.

For the larger surface units and piled units information to a greater depth is required, and recourse must be made to the range of techniques given in Fig. 2. The ROV- and submarine-based tools are again restricted in their penetration depth, but can be useful in delineating surface features, particularly for infield flowlines.

Major offshore site investigations are performed from purpose-built vessels which can be thought of as miniature versions of the drill ships deployed by the oil industry. Borings are performed using standard oil industry practice, normally in 5 in.* API pipe. Sensitive heave compensation designed to fix the drill pipe during sampling and testing has recently been supplemented by sea-bed clamping devices. Modern practice entails sampling using hydraulically operated wireline samplers in advance of the open drill bit.

In view of the inevitable disturbance caused to a sample by its removal from its virgin state, *in situ* tests have been developed to supplement laboratory tests or recovered samples. The most well known is the cone penetration test mentioned earlier. This tool can be deployed by wireless ahead of a drill bit, or can be operated from one of a range of sea-bed units.

In extolling the virtues of acquiring site-specific data it is perhaps useful to give an indication of the costs involved. The daily rate for geophysical vessels lies in the range £3000–6000, depending upon the scope. The daily rate for site-investigation vessels with all the necessary support services (positioning, drilling, etc.) is currently in the range £13 000–20 000. Rates naturally depend upon whether dynamic positioning is required, and are also subject to some seasonal variations.

In the absence of any soil data at a site the survey should consist of the following:

- mudmats: one borehole to a depth equal to the least dimension of the mudmat; cone penetration tests at 20 m spacing or less to a depth of about 5 m;
- piled foundations: two boreholes to a depth of at least 30 m; cone penetration tests at each corner and at the centre of the plan area of the unit.

*1 in. ≈ 2.54 cm.

The above provide guidelines only, and it is imperative that preliminary design be performed during the investigation by the supervising geotechnical engineer to check on the adequacy of the data.

The site investigation can also be used to obtain ancillary data for design. Probes which can be incorporated within the cone penetrometer include devices for measuring electrical and thermal conductivity. The results can, respectively, be used in the design of cathodic protection and in the analysis of thermal aspects of wells and pipelines.

A seabed survey by television camera is often performed at the same time as the site investigation by mounting a camera on the sea-bed frame. Such a survey would show up any boulders on the surface that are likely to cause later levelling problems.

CONCLUSIONS

This chapter has outlined the range of ways in which the geotechnical behaviour of the sea bed can influence a subsea installation. They clearly highlight the need for site-specific data at an early stage in the development. They also demonstrate the importance of ensuring that the operators of such installations are aware of the design assumptions made, and of the possible implications of their actions on those assumptions.

REFERENCES

American Petroleum Institute 1984. *Recommended Practice for Planning, Designing and Constructing Fixed Offshore Platforms.* API RP 2A (15th Edn).

Tjelta, T. I., Guttormsen, T. R. and Hermstad, J. 1986. Large scale penetration test at a deepwater site. *Offshore Technol. Conf., Houston, 1986,* Paper OTC 5103.

10

Developments in subsea piling

R. Martin, Anchortech Ltd

INTRODUCTION

This chapter is concerned with the suitability of long and wide piles for different subsea applications. A brief overview of the development of large hammers for driving long jacket piles is presented, and the suitability of wide piles for other applications is discussed. Methods of interconnecting wide piles are described, and finally there is an account of the development and testing of a launching system for piling off small vessels.

Note: In the context of this chapter, a "long" pile has an aspect ratio (length : breadth) in excess of 20 : 1, whereas a "wide" pile has an aspect ratio of nearer 5 : 1.

APPLICATIONS FOR OFFSHORE PILES

Applications for offshore piles may be divided into three broad categories.

- Jacket piles to support production platforms. Pile loads are principally axial compressive, supporting the structural weight,

Advances in Underwater Technology, Ocean Science and Offshore Engineering, Volume 11 – Underwater Construction: Development and Potential

with some lateral and tensile loading due to the overturning effect of storm conditions on the platform.

- Subsea structures such as templates and protective structures. Pile loads are much smaller than for jacket piles, principal loads being horizontal or vertical, compressive or tensile, caused by snagging or deflection of fishing gear. Self-weight compressive loads are generally small by comparison.
- Anchor piles for permanent moorings of floating production systems, loading buoys and circumstances where sea-bed flowlines do not permit conventional moorings. Pile loads are principally horizontal, with some vertical tensile in extreme conditions.

Since engineering solutions are predicated by available techniques and the mainstream of offshore hammer developments has been towards large hammers for jacket piles, a brief overview is appropriate.

OVERVIEW

Mainstream piling techniques

The two principal techniques for installing piles offshore are (a) impact driving and (b) drilling and grouting. Other techniques such as jetting, vibration or jacking may be advantageous or of assistance in certain circumstances, but are of application to a more limited number of soil types, and are unreliable in others. Offshore construction projects tend to be of much shorter duration than their onshore equivalents, employing more-expensive equipment. The penalties of downtime are therefore more severe, so that driven or drilled piles have been most frequently used.

The principal application for offshore piling has been to support the weight of steel jackets and production platforms on sedimentary soils. Bearing capacities up to 2000 tonnes/pile may be required, with typical pile diameters of 1–2 m to transmit the load.

Driven piles are better suited to such large-scale applications, because the reaction to the energy input is provided by the hammer itself, and because final pile sets provide a recognized measure of the capacity of the installed pile.

By contrast, a drilling technique necessitates satisfactory torque reaction, and this presents problems at large scale. In addition, the installed grout envelope around the pile is often irregular and unpredictable in shape, so that the pile must be proof-tested *in situ* unless it is of very conservative design.

Drilled piles will be used in rocky sea-bed conditions where piles cannot be driven, and may also be advantageous when a drilling rig is already on location. However, the sedimentary soil conditions of the North Sea and the suitability of heavy-lift barges for jacket installation have combined to secure most of the work for driven piles.

Hammer design

Early offshore structures in shallow water were piled with conventional hammers – generally steam or diesel – and connected to the structure above the waterline. Deeper water necessitated the use of followers as the hammer could not be submerged.

The progress of offshore exploration into deeper water, particularly in the North Sea, coincided with the development by various manufacturers of hydraulic hammers for use onshore with less environmental impact than their steam or diesel counterparts. The possibilities of a submersible hydraulic hammer were quickly realized, and had been proved by the late-1970s with the installation of piles in over 1000 ft* water depth in the Gulf of Mexico.

The adoption of hydraulic hammers was not eliminated the problems of power transmission from the prime mover above the water-line to the pile in deep water: rather, energy that was transmitted through the follower under the steam hammer is now transmitted by umbilical to the hammer directly on top of the pile. Whereas early hydraulic hammers were powered by hydraulic umbilical from the surface, the size of hammers is now such (up to 500 tonnes in weight) that this is impractical in terms of power losses and umbilical size, so power transmission is now by high-voltage electricity to a subsea hudraulic power pack coupled directly into the hammer.

Current practice in driving jacket piles offshore involves final driving by hydraulic hammer in almost all circumstances. Initial driving may be by steam or hydraulic hammers. Hydraulic hammers have been further developed into a slim format, enabling them to pass through the pile guides on the jacket, thereby minimizing pile/follower length.

Summary

The slim, electrically driven hydraulic hammer therefore represents

*1 ft ≈ 30.5 cm.

the state-of-the-art solution to piling offshore jackets, and hammer capacities (and barge lifting capacity) have reached a level to satisfy all foreseeable needs. Indeed, if the lower oil price encourages the use of smaller platforms, then further development of large subsea hammers may not occur for a considerable time.

INSTALLATION TECHNIQUES

Impact-driven piles must be supported at the desired inclination until pile penetration is sufficient to support the hammer weight on top of the pile, and to prevent wander. Pile length and hammer weight both influence instability, which is clearly greatest for long piles driven by heavy hammers, such as for jacket-bearing piles. Laterally loaded piles can be shorter, and if no greater bearing capacity is required hammers can be lighter. Pile stability is therefore much easier to maintain.

Jacket piles are supported during driving by conductors on the jacket itself. These add considerably to the weight of the jacket, but since they are also required for structural connection between the piles and the jacket, this is no great disadvantage. Similarly, the large craneage required for pile handling will be on site anyway for setting the jacket.

If the pile conductors are to be eliminated from the structure, or indeed if isolated piles are to be driven, then other means of support are required. One of the most simple in concept is a free-standing support structure in which the pile and hammer are mounted (Fig. 1). The pile and hammer can then be placed on the sea bed without the installation vessel maintaining exact station during piling operations.

In practice the free-standing piling rig is only applicable to short piles: for long piles the rig would need to be so massive that it would create as many handling problems as it solved. Nevertheless, there are still a number of applications where short piles can be driven from a relatively small vessel using a compact submersible piling rig: isolated pile anchors, where wide piles are technically preferable to long piles, and lightweight templates that do not require heavy-lift vessels for installation.

LATERAL-RESTRAINT PILES

For subsea structures requiring lateral restaint, long tubular piles have often been used because the technology is developed rather

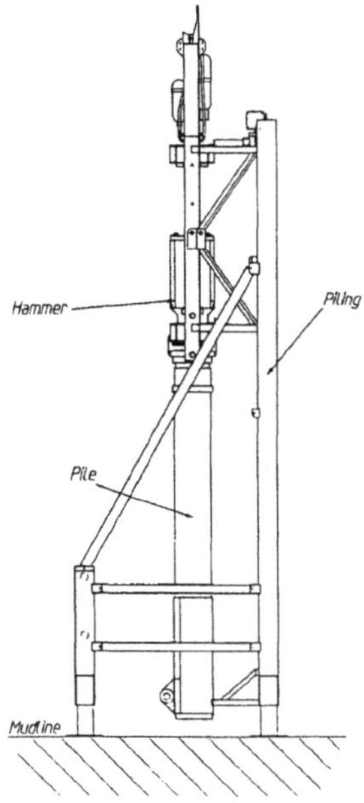

Fig. 1.

than because of any overriding technical merit. Indeed, a long slim pile derives its fixity in the soil at much greater depth than a short wide pile, so greater movement under load can be anticipated at the structure. A slim pile is also more flexible than a wide pile, and this contributes further to structural movement.

When lateral-restraint piles are used as anchors they are far more efficient when the chain is attached towards the middle of the pile rather than at its top end on the mudline. A wide pile will typically restrain twice its mudline capacity when loaded at the centre of soil pressure, and for long piles the effect is even more marked.

To achieve this higher installed capacity, part of the chain down to the padeye will be permanently below the mudline, and therefore cannot be inspected. This operational drawback, common to both long and wide piles, can be alleviated by good catenary design, but the problems are more severe for the long pile with its deeper padeye. Further, if the padeye is very deep, then the chain angle at the padeye may be nearly vertical, even under horizontal chain tension. This raises the possibility of vertical pile failure under horizontal load.

Wide piles therefore exhibit a number of advantages in restraining lateral loads compared with long piles, which are better suited to vertical bearing restraint. Given similar installation techniques and costs, the wide pile will be selected for pile anchors and the long pile for jacket foundations.

For subsea structures requiring vertical and horizontal restraint the choice would depend upon the relative magnitudes of vertical and horizontal restraint. In the case of lightweight manifolds and wellhead-protection structures, the vertical deadweight loads are trivial compared with the horizontal loads imposed by fishing gear, although in the case of tall wellhead structures overturning moments under horizontal load can generate substantial vertical loads as well. Nevertheless, wide piles can be advantageous, particularly when installation techniques are taken into account.

Examples of wide piles

A wide pile for lateral restraint can simply be a large-diameter tube. However, since horizontal load is generally applied in a single direction, the pile width can efficiently be increased in the plane only at right angles to the direction of load.

An example of such a wide pile is shown in Fig. 2. In this case a tubular pile 6.0 m long by 24 in.* diameter has its width increased to 2.0 m by a triangular thrust box near the mudline. The thrust box mobilizes much greater soil restraint close to the mudline, with reduced bending moments in the pile itself and improved rigidity under load. Chain attachment is by padeye to the bottom of the thrust box, and in this instance the portion of chain permanently buried would be oversized and designed as part of the anchorpile.

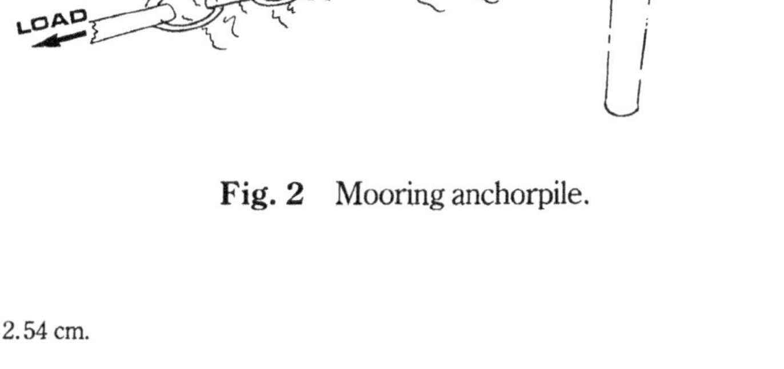

Fig. 2 Mooring anchorpile.

*1 in. ≈ 2.54 cm.

Wide piles were used to "staple" a lightweight protective structure to the sea bed for Shell (UK) Expro (Fig. 3). A crook was fabricated into the top of the pile, and the location of this crook over the perimeter tube of the cover provided the structural connection. (In this instance the sea-bed material was stiff clay, so no bearing restraint was required for the piles.)

The perimeter-pile concept enabled the structural weight to be minimized, thereby minimizing the risk of the pipeline during placement of the structure itself. Piling was performed by a 5 ton * hydraulic hammer mounted in a free-standing piling rig which was lowered down guide wires to the structure. Pile conductors were therefore eliminated, with consequent savings in structural weight and overall size.

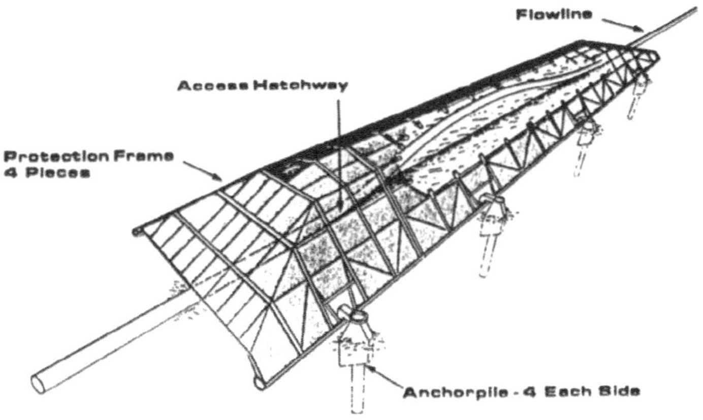

Fig. 3 Flowline protection frame.

This work indicated that turnaround times of 2 h/pile can realistically be achieved in 150 m water depth (1 h 35 min was achieved on two occasions). In cases of this type, therefore, equipment mobilization/demobilization time greatly exceeds working time, so there is no great cost penalty in installing a larger number of piles. If so doing enables a smaller installation vessel to be used, then substantial cost savings can result. Conversely, if a heavy-lift vessel is required for installation of the structure, then little benefit accrues from these techniques.

"Crook-Latch"

A limitation of the simple crook connection is that while horizontal and vertical tensile loads are transmitted into the pile, vertical

*1 ton ≈ 1.016 tonne.

compressive loads such as structural weight are not restrained, and in such conditions the structure will tend to fall out of the crook. This may be remedied by incorporating a latch in the top of the pile (Fig. 4), and the resulting "Crook-Latch" constitutes a full pin-jointed structural connection that is an automatic consequence of driving the pile. The integrity of the connection can be simply inspected visually without specialist tools.

Crook-Latch is not seen as an immediate alternative to grouted or swaged connections for jacket-bearing piles: the loads are too large

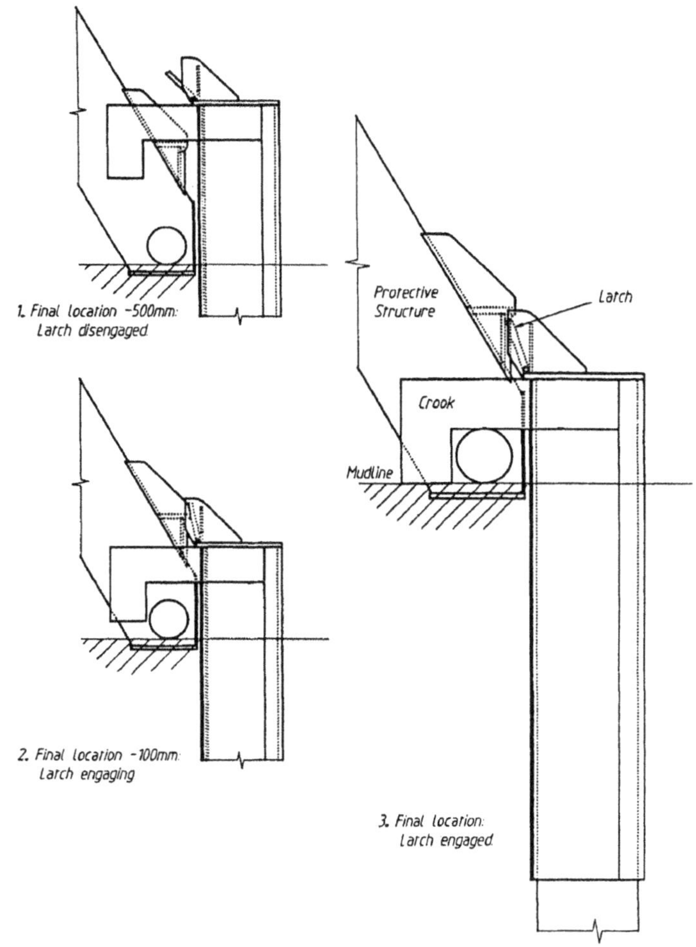

Fig. 4 "Crook-Latch" (patent pending), a structural pile connection requiring no retro-fit activities after piling. Used on perimeter piles round a protective structure it transfers full horizontal and vertical restraint (uplift and bearing) into the pile.

and the pinned connection too flexible to be practical on such tall structures. It is, however, a highly cost-effective alternative for connecting subsea structures to wide piles, where dead loads are small compared with the live loads which dominate the pile design. The flexibility of the pinned connection is of less consequence on a compact structure, and this is offset by the greater rigidity of the wide pile, and by piles positioned to maximum advantage around the perimeter of the structure.

Pile clusters

The speed with which a submersible piling rig can operate (2–3 h turnaround per pile) indicates that mobilization/demobilization time will be greater than piling time in most circumstances. As an example eight piles could be driven per day, but 4 days more would be required for loadout and steaming to and from a northern North Sea location, giving 5 days of total vessel time. Driving 16 smaller piles for the same purpose would take 6 days of vessel time, a 20% increase. But if in so doing a smaller vessel were used with a 50% saving on day-rate, then substantial cost savings would accrue. Hence the economics of driving a cluster of small piles to restrain a large load.

An example of this was the driving of four 120 tonne piles through a spreader frame to achieve a 480 tonne flowline strongpoint (Fig. 5).

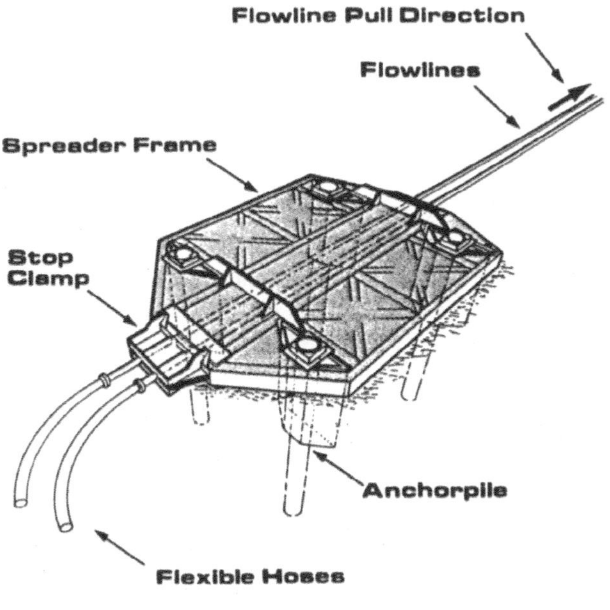

Fig. 5 Piled flowline anchor.

Symmetry across the flowline dictated an even number of piles, and doubling the number of piles from two to four did not double the total installation time on location. The benefits in this instance were smaller piling equipment and simpler handling. The piling rig was run down guide wires to the frame, so diver intervention was required to connect the guide wires.

One codicil to pile clusters is that they should not be so tight as to generate interference effects between the piles and weaken the group. This constraint may make an interconnecting framework unattractively large.

"Arrester Mooring"

As an alternative to pile clusters, small piles may be coupled together for marine applications in a similar manner to an arrester wire (Fig. 6). A chain is pinned at both ends to the sea bed by pile anchors, and hook-up is achieved by picking up the chain with a suitable anchor or

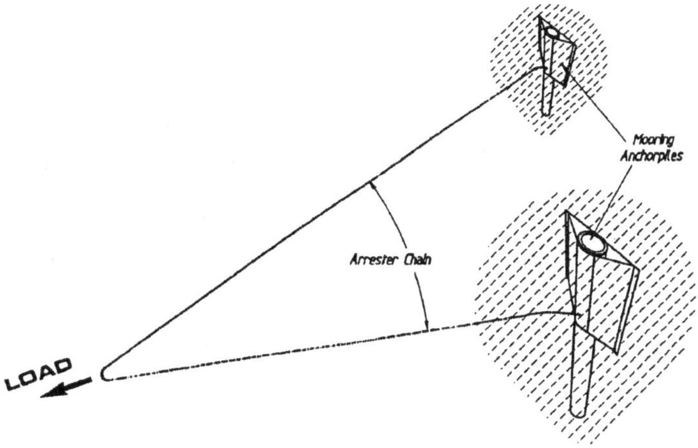

Fig. 6 "Arrester Mooring" (patent pending). A permanent mooring capability, featuring:
1. Hook-up by anchor or grapnel: no surface buoy requirement.
2. No movement under load: ideal near seabed flowlines.
3. Shorter catenaries.
4. Large loads in poor soils; small Anchorpiles.
5. Installation from small vessels; no diving requirement.
6. Disconnection by pennant wire or chaser.

grapnel. Pile position is not critical, and with chain length greater than the water depth, diverless installation is possible. As with pile clusters, applied load is distributed more evenly into the sea bed than by a single pile, and the "Arrester Mooring" offers diverless connect/disconnect capability without the need for a surface buoy and other mooring tackle in the weather zone.

DEVELOPMENT OF A PILING-RIG LAUNCHER

With the range of applications for wide piles as described above and the successful operation of a submersible piling rig off DSVs with adequate cranage, a launcher to handle the piling rig off vessels without craneage was developed.

Design parameters

The following design parameters were established:

- operation off supply boats and anchor-handlers available throughout UK North Sea, in up to 200 m water depth;
- simple onshore mobilization/demobilization to port;
- ease of loadout on to a range of vessels without modification;
- minimal lift height for stability;
- ROV intervention only;
- rotational control during lowering to sea bed;
- capable of launching a piling rig 13 m high;
- launching system under sole control of piling crew.

An A-frame concept was chosen for maximum stability on the supply boat; this would also assist in handling the piling rig over the stern roller, thereby reducing the lift height. But a conventional A-frame would need to be sufficiently wide to allow the piling rig to pass through, making it unacceptably wide for road transport. Similarly, it would need to be very tall and pivoted well inboard if it were to recover the piling rig inboard. Moreover, a pivot well back from the deck edge could impose large point loads on an unstiffened area of the deck, and would therefore seriously restrict the choice of vessels.

A sliding A-frame concept therefore emerged (Figs 7–12). This could pivot at the deck edge or stern roller, thereby transmitting vertical loads into the side of the vessel, and yet be capable of sliding inboard on rails to recover the piling rig inboard. The A-frame concept was also combined with a skid concept for pulling the rig inboard. This enabled the A-frame width to be reduced to the piling-rig width, since the piling rig never passes through the A-frame. Similarly, it enabled the A-frame skid to be cross-braced, thereby considerably reducing the weight of construction.

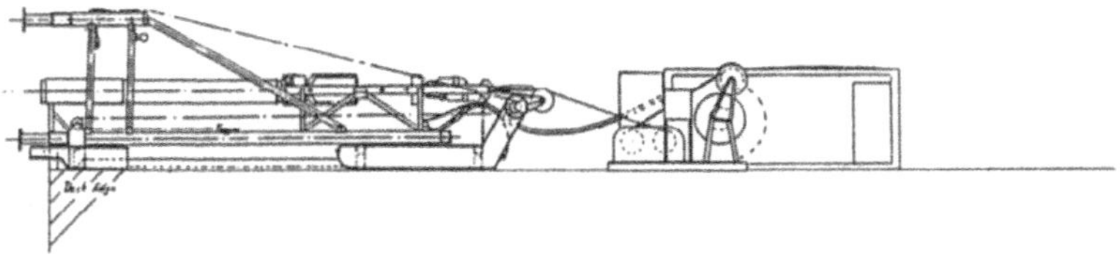

Fig. 7 Launching sequence (1).

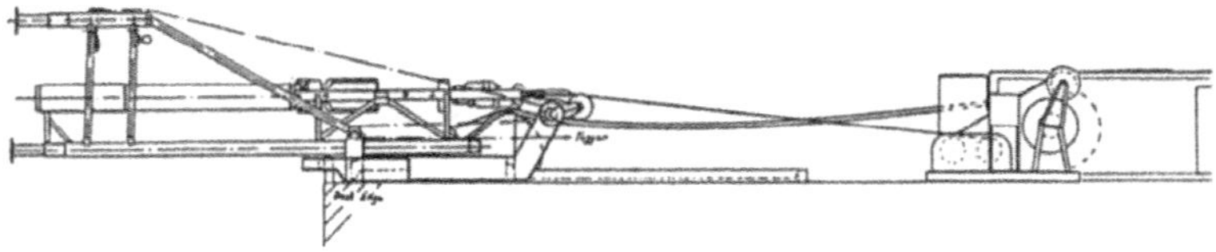

Fig. 8 Launching sequence (2).

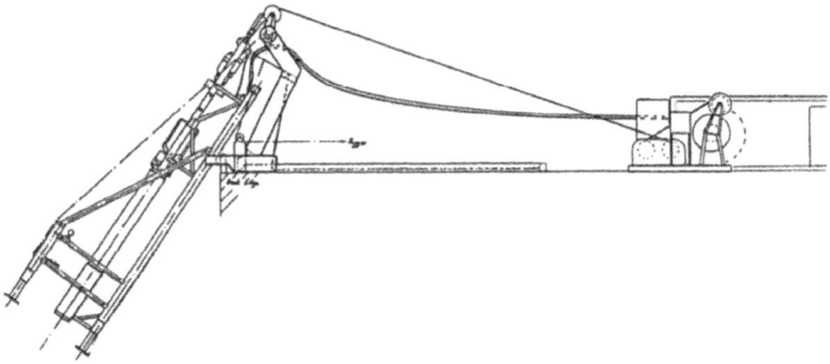

Fig. 9 Launching sequence (3).

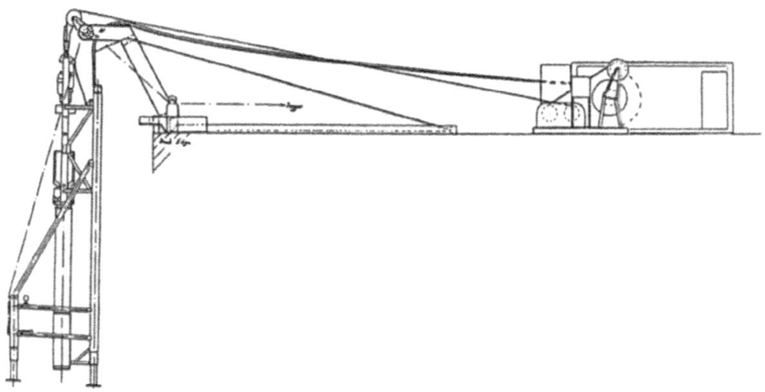

Fig. 10 Launching sequence (4).

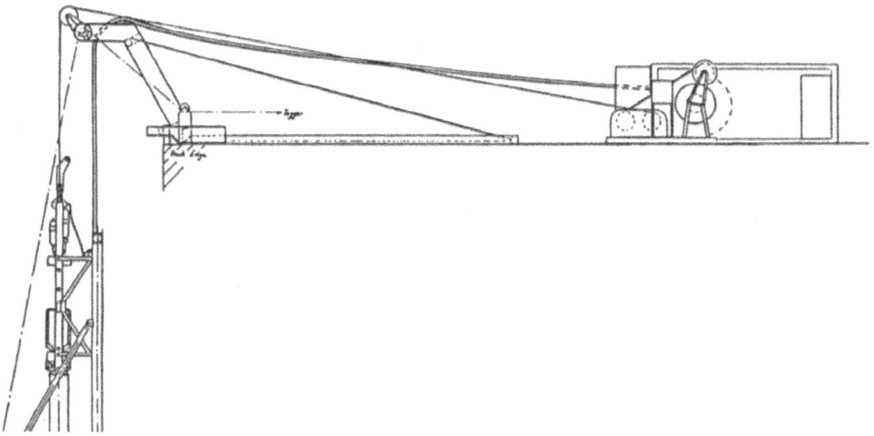

Fig. 11 Launching sequence (5).

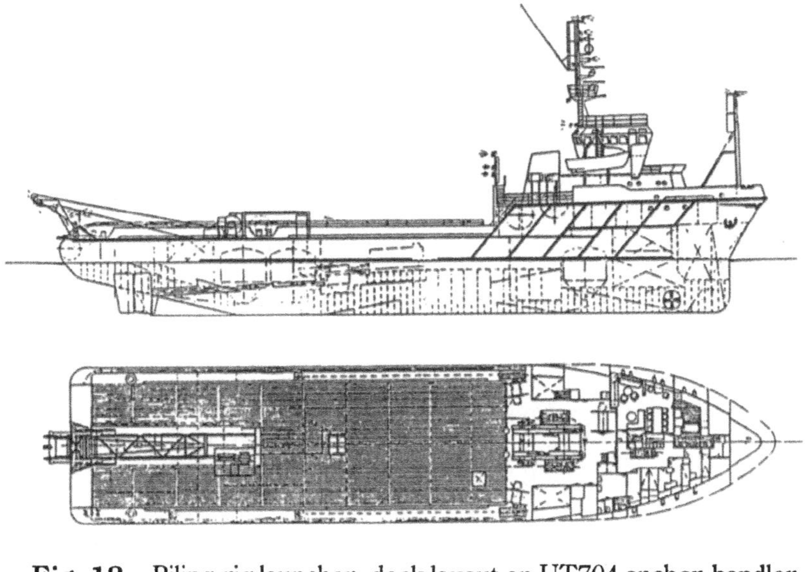

Fig. 12 Piling-rig launcher, deck layout on UT704 anchor-handler.

The means of powering the A-frame were considered. Hydraulic rams were rejected because of the length of stroke required (over 6 m) and because of their associated point loads on the deck. A two-winch system was therefore developed, consisting of a 16 tonne main capstan winch and a 5 tonne tugger winch: the tugger winch would tend to launch the rig, while the main winch would recover it. In practice the tugger would be set to constant tension during launch and recovery, with the main winch paying out to launch or pulling against the tugger to recover. Use would be made of the vessel's tugger winch for this purpose. The main winch would lower the piling rig to the sea bed.

The weight of the piling rig and pile at 15 tonnes is such that it cannot be lifted by its umbilical; lift wires and umbilical must therefore be kept separate. The umbilical is maintained in constant tension (about 1 tonne) to the piling rig, and care must be taken to ensure that the piling rig does not spin and wind the umbilical round the lift wire. The main lift wire was therefore double-reeved at 2.0 m centres to give good directional control during lowering. Additionally, this doubled lifting capability to 32 tonnes. Similarly, the tugger winch was doubled-reeved to provide 10 tonnes of launching force. Consideration was given to using two single-reeved tuggers to achieve the same end, but this was rejected on the grounds of difficulty in co-ordinating the ship's tuggers.

In its deployed position the A-frame would be restrained on static wires just past "top-dead-centre". This would ensure that the A-frame did not fly back inboard under winch tension, while mimimizing point loads to the deck from the static wires. Tie-down strops were also fitted as a precaution. The piling rig was pinned to the top of the A-frame during launch and recovery, this pin being removed hydraulically after connection of the tie-down strops.

The static wires from the centreline of the rails to the edges of the A-frame provided good lateral stability to the A-frame under rolling conditions. Care was also taken to ensure flexibility between the piling rig and the A-frame, so that rolling movements of the piling rig would not break the A-frame out of its supports. Catchers and guides were provided on the rails to centralize the piling rig into the A-frame during recovery.

The system therefore consisted of the piling rig with A-frame skid and rails, the main wince, and a 20 ft container with umbilical and piling controls. The piling rig with A-frame skid and rails was loaded out as a single lift on a 40 ft lorry, and at 3.2 m wide can be moved on UK roads without police escort. (The front legs of the rig are folded down to reduce the load height to 2.8 m.) This was then placed on the vessel centreline at the stern roller in a single lift. The main winch and 20 ft container, mobilized on a second lorry, were set out on deck with respect to the rails. Structural connections between the winch and the rails were provided to mimimize sea-fastening, and these were made by tensioning the main winch to tie all of the components into each other before sea-fastening the winch and container.

The main sea-fastening for the system occurs at the deck-edge/stern roller where 8 in. tubulars are welded in position. The rail system is cross-braced and gives great lateral stiffness along the length of the structure, so that only minor structural connections are required at the inboard end of the rails. Once tied in, the winch and container require only nominal sea-fastening, and on a UT704, all such connections can be made directly to the steel deck area so that no timbers need to be removed.

Offshore trial – piling-rig launcher

A trial of this piling-rig launcher was conducted early in 1987 off the UT704 anchor-handler *British Piper*. The extent of onshore trials beforehand was necessarily limited because of the requirement for 7 m of water depth at the stern roller (or nearly 10 m under the launcher) to deploy the piling rig fully.

Initial trials in Edinburgh Dock were performed using the vessel's pennant wire and tugger winches in turn to initiate the launch. The pennant drum could not generate enough tension, while the tugger generated too much. In both cases they had difficulty in coping with slow launching speeds for the first trial, and tended to stall. In the short term a hydraulic compensator was introduced into the wire to maintain 5 tonnes of tension against the tugger winch; in the long term a tugger would be incorporated into the piling system itself. Welded connections were introduced to stiffen the sheave assembly on the piling rig, but after these modifications the equipment was launched and recovered satisfactorily.

Launching was also performed in the Firth of Forth in Seastate 3. Operations were hampered by sheave lead angles being greater than anticipated, but otherwise the system showed itself operable in moderate open-water conditions.

Apart from the technical matters mentioned above, the following operational points also emerged.

- ROVs, while standard equipment on DSVs, are a major extra item for short-piling contracts on supply vessels in terms of transport, sea-fastening, operating personnel and deck space. Ultimately the piling rig must be completely self-sufficient in observation and control.
- Power hook-up on load-out must be planned well in advance to minimize costly delays.
- The main propellers should not be driven at full power while the piling rig is held in the A-frame. Launch procedure should therefore be to head into the weather and idle during the launch, only manoeuvring into position once the piling rig is well below the propellers, say at 20 m depth.
- On-board accommodation is limited, and with no helideck this

imposes additional constraints on the work possible from this type of vessel.

- The trials indicated a launch time of 2 m, and 15 m/min lift/lowering speed. Thirty minutes would therefore be required for full launch and recovery in 150 m of water depth, which is well within the time-frame of 2–3 h/pile turnaround time.

CONCLUSIONS

Although mainstream piling development has been associated with long bearing piles for offshore jackets, and this work will continue, other applications associated with subsea structures and floating production systems are better suited to wide piles which can be installed off smaller vessels.

Structural connections can be incorporated into the design of the wide pile, and small piles can often be economically coupled together to restrain greater loads. Finally, piling can be performed off vessels without craneage, such as anchor-handlers, with rapid-mobilization piling equipment.

11

The role of heavy-lift vessels

J. P. Mitchell, Micoperi Ltd

INTRODUCTION

The continued development of offshore installation equipment has allowed new methods and techniques to be employed for offshore installation. This is true for both subsea developments and conventional platforms, and has enabled large-scale underwater construction to be carried out more efficiently. This chapter looks at the development trends of offshore construction vessels and their effect on subsea installations, and expands on some of the areas where heavy-lift vessels have an important role to play.

INSTALLATION VESSEL DEVELOPMENT

When the search for hydrocarbon resources first moved offshore, the floating construction equipment necessary for installation consisted mainly of flat-bottomed barges with relatively small cranes. This was all that was required for the initial development areas of the US Gulf, and similar mild environs. When more-hostile areas, such as the southern and central North Sea, were developed ship-shaped vessels of larger crane capacity proved far more capable and quickly

Advances in Underwater Technology, Ocean Science and Offshore Engineering, Volume 11 – Underwater Construction: Development and Potential

emerged as the minimum requirement for working at such locations. Crane vessel lift capacity soon increased from less than 800 tons* to over 2000 tons, as it became obvious that the bigger vessels with greater capability were the most cost-effective method of installation.

However, when very hostile waters such as the northern North Sea were encountered even the more sophisticated monohull crane vessels found it took a whole season to complete a major installation. As operators could not afford these project delays, the industry started looking for ways of reducing installation times, extending workable weather windows and enlarging the scale of offshore operations. The semi-submersible hull form, long known as one of the most stable platforms from which to operate in hostile environments, combined with dynamic ballasting systems, proved to be ideal for heavy-crane lifting operations in harsh conditions.

Workability improved dramatically, giving the possibility for the majority of offshore tasks to be undertaken year-round (Fig. 1) and allowed more-efficient subsea operations to be performed, especially when remote or diverless operations were employed.

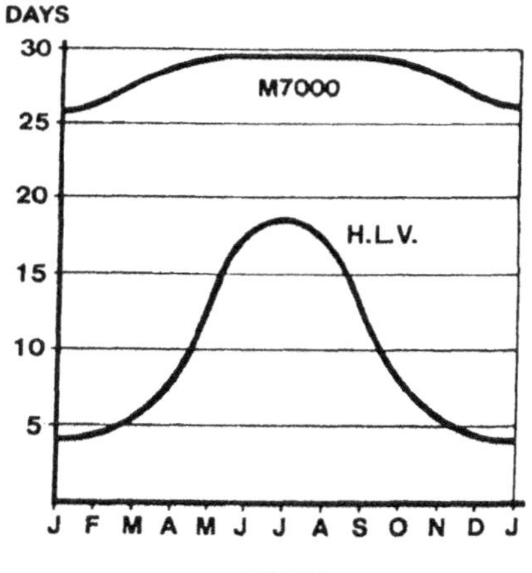

Fig. 1 Comparative workability (northern North Sea).

Individual offshore crane capacities have increased tenfold in the past 15 years, with total vessel lifting capacities growing from 700 tonnes on monohull ships to 14 000 tonnes on SSCVs during this period (Fig. 2).

Whilst this increase in capacity was aimed mainly at jacket and module installations, allied with it was the ability to maintain station

*1 ton ≈ 1.016 tonne.

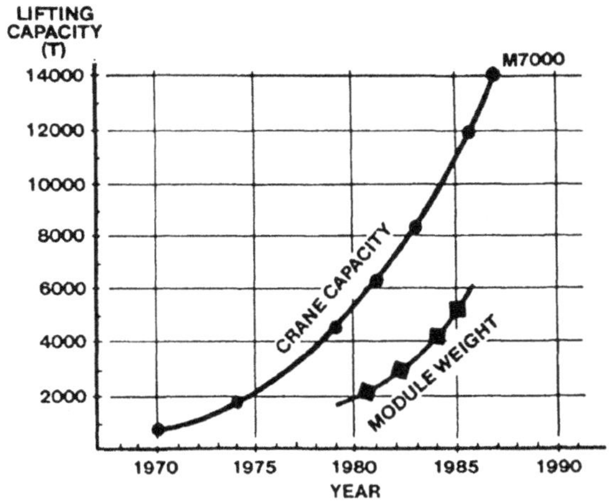

Fig. 2 Crane capacity trend.

accurately, to provide a very stable platform for offshore operations and to operate a large range of equipment designed specifically for deep underwater operations such as large-capacity subsea hoisting blocks, high-energy hydraulic underwater hammers and deep-dive systems. These have proved invaluable for a number of underwater constructions, and most subsea templates and manifolds over 300 tons in weight have been installed with heavy-lift vessels (HLVs).

HLV INVOLVEMENT IN UNDERWATER CONSTRUCTION

A certain amount of subsea construction has been associated with most field developments and has usually involved HLV operations. This ranges from the installation of subsea items associated with conventional platforms, such as pipeline tie-ins, to construction of complete subsea production systems. Some examples of where HLVs have been used for underwater construction are listed below, the list indicating their range of involvement:

- remote anode installation;
- pipeline tie-ins;
- multiwell drilling template installations;
- manifold centres and valve-station installations;
- docking-pile installation;
- pipeline stabilization;
- platform repairs;
- TLP foundation installation;

- anchor installations for loading buoys/guyed towers;
- supply-boat mooring base installation;
- sea-bed production systems installation.

The unique capabilities of HLVs were required for all of the above examples, and some of these operations would not have been possible without the higher-capacity SSCVs only recently available. The latest generation of crane vessels, with even better capability, offer designers the possibility to engineer more cost-effective subsea solutions to some marginal-field problems.

HLV ADVANTAGES FOR SUBSEA ACTIVITIES

There are a number of HLVs currently available, covering a wide range of hull forms and lifting capability. As it is difficult to include the details of all of these vessels, we have considered only one type of HLV – the semi-submersible twin-crane vessel. For the purposes of this chapter we have chosen the SSCV Micoperi 7000 because it embodies the basic capabilities of all other HLVs and is able to demonstrate how the latest technology featured on this vessel can affect today's underwater construction tasks.

Heavy-lift vessels have a number of advantages over smaller installation craft for subsea activities, and these advantages primarily lie in their ability to provide a very stable platform from which to carry out subsea operations, heavy lifting and lowering capability, and a complete range of equipment for efficient construction and installation. The advantages of HLVs can be demonstrated by looking at the particular capabilities of our chosen vessel in such areas as:

- station keeping;
- operability;
- lifting capacity;
- subsea installation equipment;
- associated facilities;
- deepwater pipelaying.

Station keeping

By virtue of its mass and efficient semi-submersible configuration, the vessel has good sea-keeping characteristics and with 16 anchor wires, each over 3 km in length, the vessel is able to moor out in water depths of up to 450 m and maintain station in the harshest environments envisaged (Fig. 3).

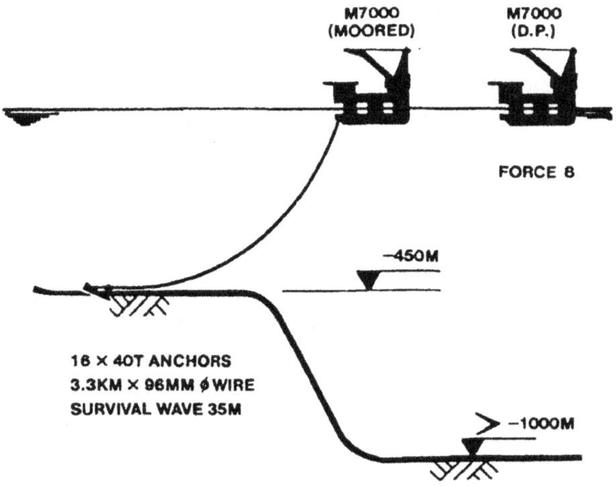

Fig. 3 Station keeping.

In very deep water, or at offshore sites with very congested pipeline configurations where anchor moorings are difficult to lay, the vessel can accurately maintain station with its dynamic positioning (DP) system referenced by either sea-bed transponders, satellite positioning or taut-wire systems. Alternatively, the DP system can complement a reduced anchor spread, the two systems working together using an integrated control system. This is ideal for short-duration jobs on existing field complexes.

The software in the DP system can be modified to incorporate the direct mooring of another vessel or structure alongside the SSCV. This is ideal, for example, for TLP tether installations when very precise positioning is required in deep water, to connect the tethers to the subsea foundation templates (Fig. 4).

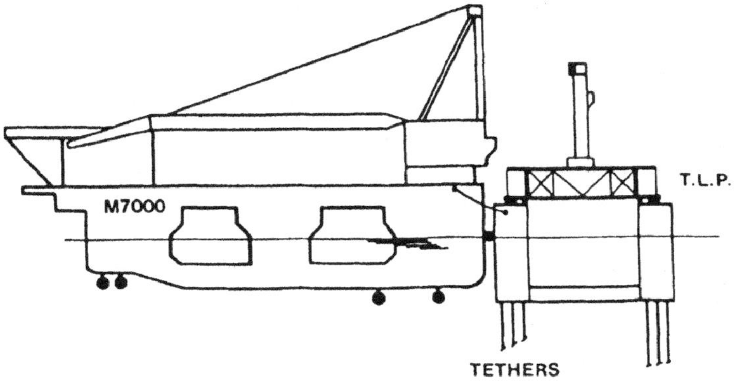

Fig. 4 TLP positioning (tether stabbing).

Operability

The optimization of the body lines, as well as good weight distribution, provides the semi-submersible vessels with low motion characteristics and leads to less weather downtime and reduced added-mass effects when lowering submerged structures.

As an example of the high operability (time working) of the SSCV, if crane-hook motions for underwater operations are limited to 1.5 m, the vessel could still work in excess of 28 days/month in the northern North Sea during the main construction season from April to September. This high operability can be restricted by support services and equipment such as diving and tug operations, which may need lower weather thresholds. However, the development of remotely activated tools and systems, together with dynamic positioning of the HLV, has led to improvements in these areas. The low vessel motions allow heavy submerged objects such as templates to be lowered to the sea bed without the need for special motion compensators or spring systems, as the added-mass effects are much reduced, and also allows setdown on to the sea bed without large impact loads.

Lifting capability

The HLV under consideration has twin bow-mounted cranes, each with the following nett lifting capacities:

Main hook	7000 tonnes at 40 m radius fully revolving
1st auxiliary hook	2500 tonnes at 75 m radius
2nd auxiliary hook	900 tonnes at 115 m radius
Whip hook	120 tonnes at 150 m radius

The nett lifting capacities (Fig. 5) include all crane tackle and allow for loadings due to wind, impact and other dynamic forces.

The first auxiliary hook is very useful for medium-sized loads, whilst the second auxiliary hook is fitted with enough cable to enable it to be used underwater at depths up to 450 m.

An example of the advantages of this lift capability can be seen when considering the installation of jacket-type structures. In the past such structures, with operating weights in excess of 2000 tonnes, have had to be launched from special launch barges. This entails building launch runners on the jacket and providing it with additional buoyancy in the form of tanks or tubes for both the launch

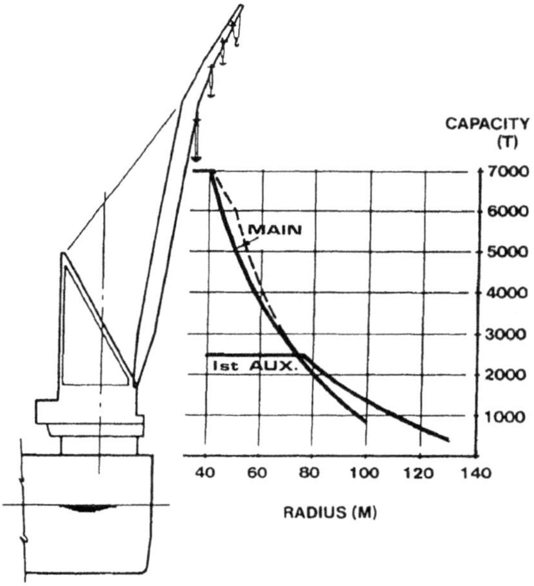

Fig. 5 Crane capacity.

and upending operation. With twin matched cranes of high capacity at a large outreach, jacket-type structures of 8000–9000 tonnes operating weight can now be lifted, saving considerably on the expensive fabrication of launch runners and buoyancy tanks (Fig. 6).

The larger hook capacities also allow for greater hook load to be applied to structures during upending, and reducing the need for complicated upending control and ballasting systems.

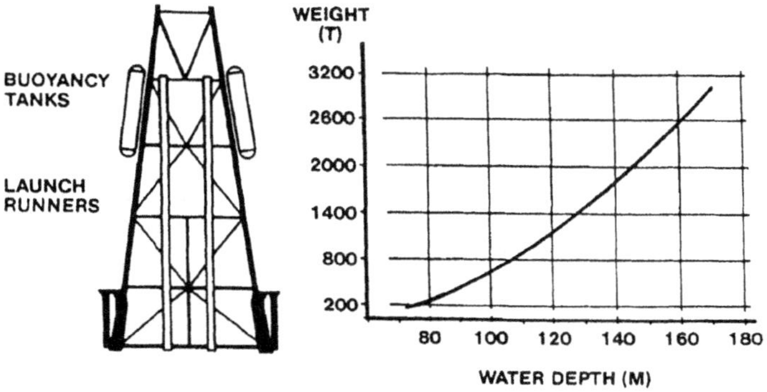

Fig. 6 Lifted jackets: weight saving.

Subsea equipment

As previously mentioned, the 900 tonne second auxiliary hoist of each crane can be used at depths of up to 450 m underwater. Alternative blocks are available, and all have been specially designed for underwater operations, and for handling underwater hammers and remotely operated lifting tools. The main and first auxiliary blocks can also be re-reeved to enable them to lower loads heavier than 1800 tonnes underwater.

The HLV is also equipped with a modular saturation-diving system for up to 14 divers, and includes a hyperbaric lifeboat and a mini-submarine powered by a closed-circuit diesel engine system (Fig. 7). The diving system is rated for 450 m water depth, and three locations for the diving package are provided on board in order to allow maximum flexibility for underwater work.

The above is all standard equipment and does not have to be specially hired or mobilized, and offers the most efficient method for installing subsea structures.

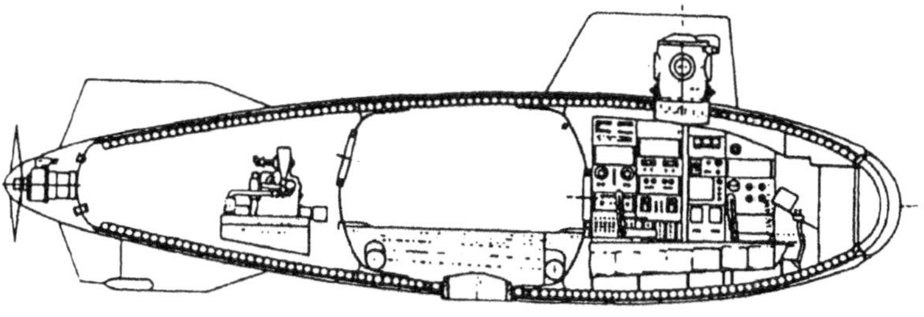

Fig. 7 Submarine CEE-22.

Associated facilities

For support of construction work the vessel has a large inventory of tools, equipment and workshops, together with a large welding plant made up of more than 50 static welding machines.

The major installation equipment includes a range of the latest Menck hydraulic pile driving hammers, including the most powerful MHU 3000 models for driving piles up to 102 in.* in diameter. The hydraulic hammers are fully equipped to allow "free-riding" or "slim-line" modes for use both above and below water. A range of hydraulically operated lifting tools with rated capacities up to 1000 tonnes are also provided.

*1 in. ≈ 2.54 cm.

The vessel can accommodate up to 800 people in the air-conditioned living quarters, mainly in one- and two-bedroom cabins, and provides other first-class welfare and recreation facilities, including a swimming pool, gymnasium and cinema.

The vessel's deck, larger than the size of two football pitches, provides ample room for storage of equipment and structures, and can transport 15 000 tonnes of cargo.

Pipelaying

The large, strong deck of the vessel can accommodate either semi-conventional multi-joint and stinger pipelaying methods or reel-laying applications. One of the reel-laying methods which has been investigated makes full use of the high specification of the vessel to lay pipelines quickly and economically in water depths up to 400 m.

The central opening in the main deck allows pipelaying from the centre of the vessel. Interchangeable reels which can accommodate large lengths of pipe can achieve a very fast lay rate, as the vessel does not need to return to shore to re-supply with pipe.

The DP system of the vessel allows pipelaying to be performed without anchoring, and the reduced vessel motions (due to good sea-keeping characteristics) and pipe entry angle allow pipelaying operations to be continued in higher seastates than previously considered.

Uneven sea-bed conditions can be accommodated more easily, reducing the cost of rock dumping or trench filling, and onshore control of pipeline production produces a higher overall quality.

SUBSEA STRUCTURES REQUIRING HLV INVOLVEMENT

The growth in lifting and placement capability has enabled large underwater structures to be installed more efficiently, and operators have been able to develop structures for subsea drilling and production without restraint on installability. A number of drilling/production templates and underwater manifolds up to 2000 tons in weight have been installed and more are currently on the drawing board. It is envisaged that these structures will require HLV involvement. These templates usually have in-built trawler and anchor drag protection, but for small single-well completions large protection structures need to be deployed, again usually by HLVs. The foundation templates for tension leg platforms are ideal candidates for HLV installation, especially as large driven piles provide the most efficient foundations. New developments currently being considered for deep water are submerged well-support towers (as proposed by Petrobras for offshore Brazil) and sea-bed production systems being considered for locations in up to 1000 m of water. In addition to these oil-related developments there are a number of civil engineering projects which could benefit from HLV involvement, including the placement of underwater caissons and tunnel sections. A few examples of HLV operations and their advantages when installing some of the above are described in the following section.

TYPICAL HLV OPERATIONS

To demonstrate their special role in underwater construction, some typical operations unique to HLVs are described below. In order to cover as wide a range of operations as possible in a few paragraphs, we consider the installation of two types of structure: a subsea drilling/foundation template and a submerged well support tower.

Template installation

The template being considered for installation could be a multiwell drilling template, TLP foundation template, subsea manifold or a similar structure. The lifting capability of the HLV enables the template to be designed without the need for buoyancy, and therefore water-depth limitations due to hydrostatic collapse are not a problem assuming all members can be free-flooding.

Loadout

The template could be loaded out on to a large supply vessel or flat-top cargo barge and brought to the HLV. A cargo barge would be more suitable if the foundation piles were transported with the structure. The large deck of the HLV is ideal for storing the template and piles, enabling all the pre-installation checks on positioning transponders and other installation equipment to be made before deployment, and therefore transfer of the template to the HLV could be advantageous. This would also allow the template to be transported offshore on the HLV, increasing the operability of the operation.

Lowering/positioning

With the lifting slings attached to the structure, it can be lifted from the vessel ready for deployment. Dependent on the natural period of the template, the sling arrangement would probably be a "rigid" system; this is made possible by the low added-mass effects, negating the need for any form of motion compensator or nylon slings.

The underwater blocks allow templates to be lowered to the sea bed in a single operation, without any sling change-overs or hang-off requirements. The high block elevations achieved through the long boom length allow for a greater length of initial slings and enable templates to be lowered to more than 500 m underwater (Fig. 8).

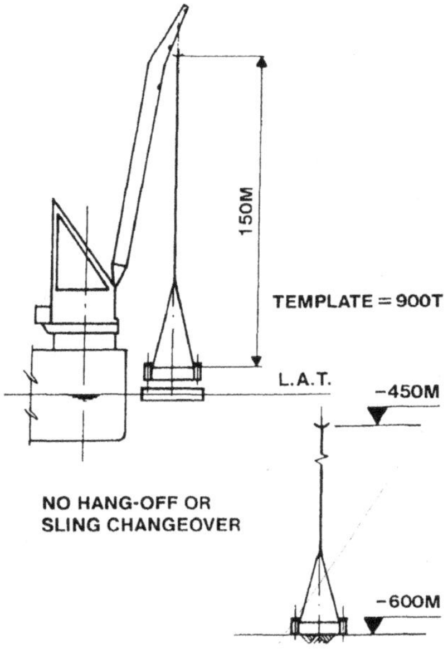

Fig. 8 Template installation.

Final position can be achieved by either crane or vessel movements, with orientation controlled by guidelines or tugger lines from the vessel or template thruster capability. For very accurate location in deep water the twin underwater blocks allow dual lowering operations to be carried out, giving better control and positioning possibilities. Re-rigging of the larger-capacity blocks also allows structures to be lowered to even greater water depths (Fig. 9).

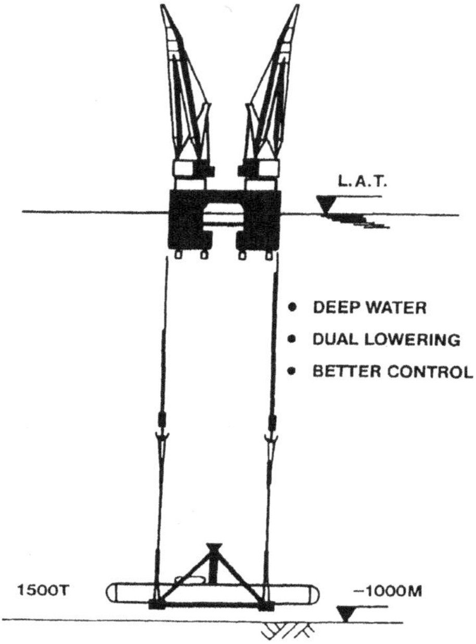

Fig. 9 Subsea installation.

Release/levelling

Hydraulically operated lifting tools which can be remotely released (either hydraulically or acoustically) have already been used on a number of deepwater template installations, allowing the rigging retrieval operation to be conducted without the use of divers. In shallower water conventional rigging arrangements released by divers could be considered. Levelling systems are very dependent upon soil conditions at the site, and mudmat/jacking systems, together with pile climbing/jacking systems have shown to cover the levelling needs of a large range of sea-bed conditions and template arrangements reliably.

Foundation supports

Certain structures, such as protection covers, can use gravity base foundations, but the majority of templates which are sensitive to sea-bed settlement, etc., require piled foundations. Driven piled foundations have proven more efficient, especially for tension loads associated with TLPs and the HLV's ability to lift and lower long large-diameter piles has proven very advantageous. The hydraulic underwater hammer spreads which form part of the standard equipment of most HLVs can quickly drive the piles to their required penetration. Support for the pile-driving system can be provided either by the template or, where this is not possible due to poor soil conditions, by the vessel/hammer backfollower system. The HLV provides a stable working platform from which the pile grouting operation can be performed or the crane can be used to deploy a pile swaging system.

Well support tower

For their Marimba field offshore Brazil, Petrobras are considering the use of a well support tower which could be used in very deep water to bring the wellheads and manifolding within the reach of divers (Fig. 10). The installation of this jacket-type structure lends itself ideally to HLV involvement.

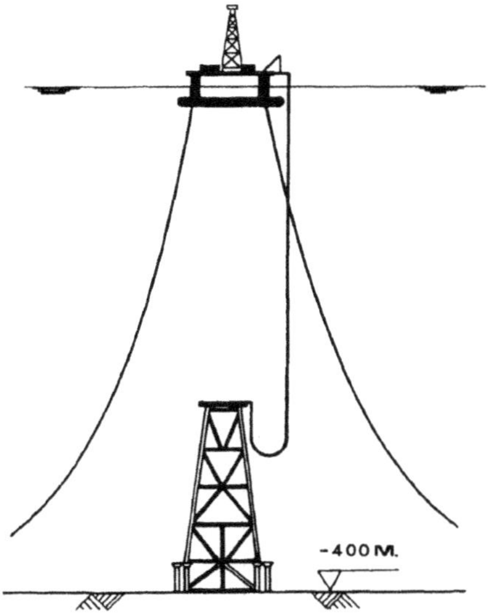

Fig. 10 Deepwater well support tower.

Loadout/transportation

The structure could be loaded out and transported on a conventional cargo/launch barge. The tower could be skidded longitudinally on to the standard skidways of the barge or, dependent upon its weight, skidded on transversely or wheeled on using a bogie system.

Lifting/lowering

With the transportation barge moored to the HLV, the twin cranes of the vessel could lift the tower clear from its barge. After removing the cargo barge, the tower could be set in the water and, by applying hook load to the top of the tower, it could be upended ready for placement on the sea bed or over a pre-installed template (Fig. 11). The advantage of this HLV procedure over a launching, ballasting and submerging operation has already been covered in the section "Lifting capability". This HLV procedure can be performed very quickly, reducing the exposure of the operation and enabling early commencement of piling, which provides the structure with better stability. Disconnection of lifting systems can be performed as for template installations.

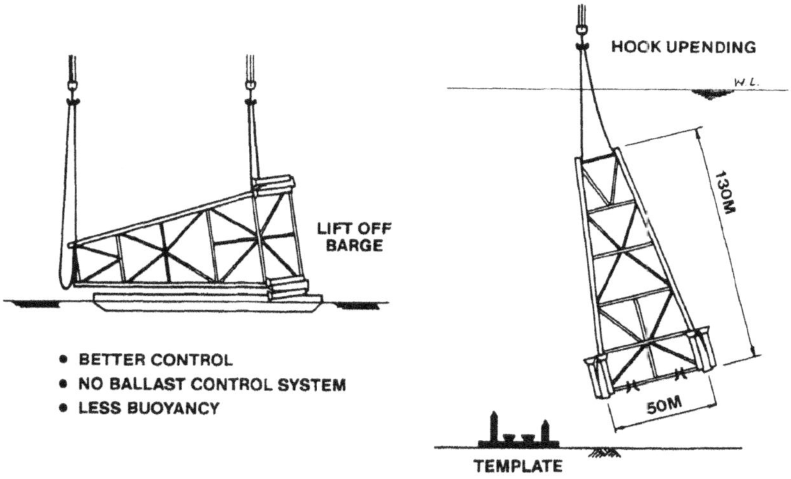

Fig. 11 Lifting and hook-assisted upending.

Piling/levelling

The foundation piles can be lifted using a remote-release lifting tool which requires no lugs or lifting attachments on the pile. They can then be stabbed into the pile sleeves, by the cranes' underwater block, using ROVs for monitoring. The underwater hammers can similarly be placed directly on the piles, driving them to their design penetrations. The reach of the cranes enables the HLV to perform all piling operations from one set-up position, reducing vessel relocation time (Fig. 12).

Pile grouting or swaging would be as for template installations. Well tower levelling, if necessary, can be performed by pile-jacking

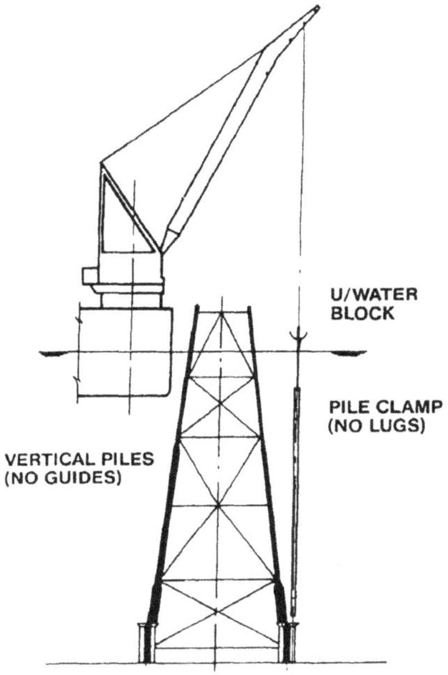

Fig. 12 Pile installation.

equipment previously used on platforms in Brazil, the Mediterranean and other locations.

SUMMARY

For heavy underwater construction work, such as the installation of large templates and manifold structures, the HLV has an important role to play. The HLV's good stability, together with its lifting, lowering and piling capabilities, provides operators with one of the most economical methods for installation of these structures, and the latest generation of heavy offshore lifting vessels enables designers to develop cost-effective subsea concepts previously not possible. The speed and reliability of offshore operations when performed by large HLVs, as opposed to less-capable vessels, provides the best security of project schedules, and allows the work to be carried out on an all-inclusive lump-sum basis.

With its overall capability and adaptability, the HLV will continue to play an important role in deep underwater installation applications.

12

Habitat for repairing submarine cables

E. G. Giles, Central Electricity Generating Board

The original electricity link between England (Dungeness) and France (Equihen) was formed by a pair of d.c. cables laid on the sea bed. The link was very unreliable due to the large number of anchors and trawl boards that damaged either or both of the cables. Repairs were continually being carried out using a specially fitted-out vessel, the *Dame Caroline Haslett,* previously a collier vessel. A typical repair is shown in Fig. 1. Once the faulty section of cable is removed a new length held in store on the repair vessel is joined to one end of the original cable.

The new length of cable is then relaid on the sea bed until the second end of the original cable is reached. This cable is then recovered to the repair vessel and joined to the new cable, the end of which is retained on the vessel. A loop of cable approximately equal to twice the water depth is finally laid on the sea bed in the form of a bight or loop. Any cable laid on the sea bed is vulnerable to anchor and trawl damage, and repair loops as described above add to the risk. Such repairs are, however, relatively easy to carry out, but subject to adverse weather.

The new electricity connection between England (Folkestone) and France (Sangatte) (Fig. 2) has the submarine cables buried 1.5 m into the sea bed. Whilst this affords protection against trawl boards and

Advances in Underwater Technology, Ocean Science and Offshore Engineering, Volume 11 – Underwater Construction: Development and Potential

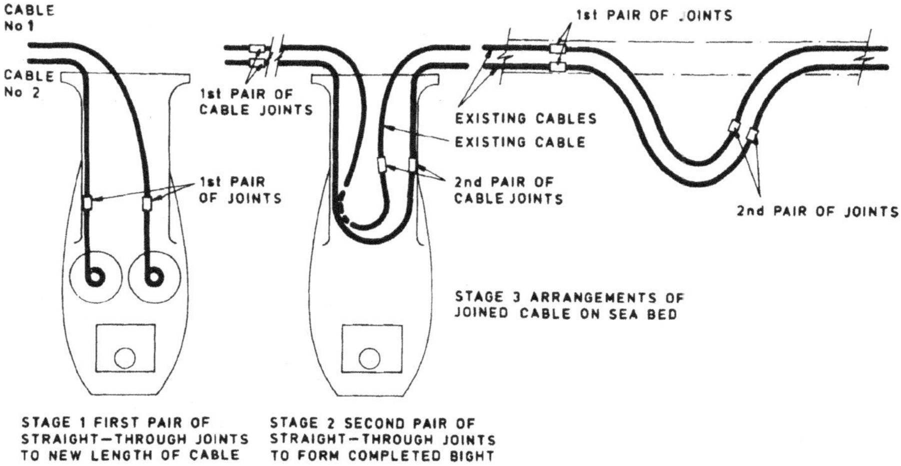

Fig. 1 A typical repair.

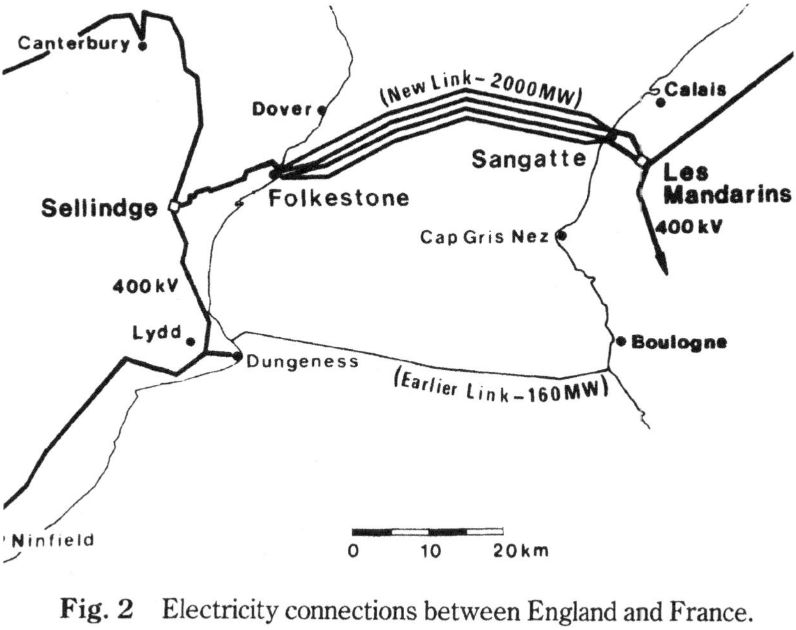

Fig. 2 Electricity connections between England and France.

the majority of anchors, it does present difficulties in repairing the cables when faults occur. If the previous method of repairs were adopted it is considered extremely difficult to bury a repair loop, and if surface protection were used the cable would be at risk where it leaves the top of a "hard" walled trench. Surface protection in many cases, i.e. concrete mattresses, is not considered a satisfactory method, although this has been accepted in certain circumstances.

To overcome the above problems of repairing submarine cables the CEGB and EdF had a feasibility study carried out. This study was to examine all the problems and determine whether it was possible to produce a suitable environment on the sea bed in which a cable joint could be made on a high-voltage cable. The prime requirement was for trained and experienced cable jointers to be able to operate in their normal environment, as it was not considered practical to train divers to carry out this work. With the age of such cable jointers being 30/35+ years it was considered necessary to have the habitat working at normal atmospheric conditions. A further reason for such conditions was that with a relative-humidity requirement of 20% with a temperature of 20°C it was not considered possible to achieve this under hyperbaric conditions.

The electricity cable laid between England and France is a conventional solid mass-impregnated, paper-insulated, lead-sheathed one with polythene oversheath and steel-wire armour, as shown in Fig. 3. In order to perform a joint between two such cables, the following procedures are carried out:

(1) the two ends are held securely and the copper conductors joined;
(2) the paper insulation is scarfed back to form a long cone;
(3) paper tapes are wound on to the two ends to reconstitute the insulation to its original diameter;
(4) the lead sheath is replaced and joined to the sheaths of the two ends;
(5) the polythene oversheath is reconstituted;
(6) the wire armouring is replaced, or the two ends joined together to give complete mechanical protection.

In order to ensure that the cable papers are not able to absorb moisture, the atmosphere in which the joint is made is controlled to a maximum of 20% relative humidity at a temperature of 18–20°C. It is important also that no air at high pressure be allowed to enter the cable; to achieve such conditions on land air-conditioning equipment is used. To achieve such conditions underwater it is fairly evident that a 1 atm (101 kPa) chamber is necessary.

Traditional cable jointing may involve the skills of brazing (the conductor), lead burning (the sheath) and welding (the armouring wires). However, in the confined area of the habitat such activities would have to be curtailed. The flame usually employed on land produces compressed inflammable gases in the habitat, which are strictly forbidden on safety grounds.

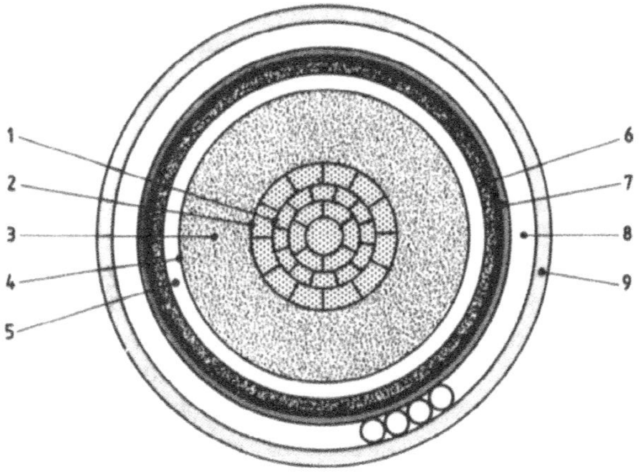

1. SEGMENTAL COPPER CONDUCTOR
2. CONDUCTOR SCREEN
3. INSULATION (MASS IMPREGNATED)
4. CORE SCREEN
5. LEAD ALLOY SHEATH
6. POLYETHYLENE ANTI - CORROSION SHEATH
7. BITUMINISED JUTE
8. 6mm GALVANISED STEEL ARMOUR WIRES
9. POLYPROPYLENE STRINGS

APPROX WEIGHT PER METRE 33.8 KILOGRAMS
OVERALL DIAMETER 104.2mm

Fig. 3 Cross section of 900 mm^2 270 KV d.c. submarine cable for cross-Channel link.

A compression type connector was, therefore, adopted for jointing the copper conductor, thus eliminating the need for a brazing torch. An electric hot-air gun or a soldering iron is used to melt the lead sheath, this replacing the traditional lead burning. The armouring is joined using a steel sleeve with mechanical grips at either end terminating the armour wires, thus eliminating the need for welding.

The paper tape insulation, in the form of rolls of impregnated paper, is carefully applied by hand and is stored in a special oven at 50°C, the rolls of tape being packed in individual containers. This eliminates the necessity for the open oil bath normally used on land which produces unacceptable gases.

The main problem area revealed during the study was the sealing of the submarine cable into the habitat. The cross section of the cable is shown in Fig. 3, and it can be readily appreciated that to seal either into the armour wires or on to the polythene or lead sheath would prove difficult when a maximum external pressure of only 100 psi (6.8 bars) could be applied to the outside of the cable. Grease and resin were initially used unsuccessfully, but use of cryogenics proved that an effective seal could be provided to withstand the maximum 6.8 bars pressure anticipated. Laboratory tests were carried out to prove the effectiveness of this seal.

With the feasibility study showing that it would be possible to provide the right environmental conditions and satisfactorily seal the cable into it, a decision was made by CEGB and EdF to proceed to the construction and testing phase of a full-size habitat. A joint venture grouping of Atelier et Chantiers de Bretagne based in Nantes and Slingsbury Engineering Ltd based in Kirbymoorside were given the contract for this work. The habitat is now constructed and has satisfactorily completed trials in the dry, submerged in a dry dock flooded to 18 m and also in the English Channel in a depth of water of 30 m.

Existing legislation in the UK required that any manned, submersible vessel must be certified as safe by the Department of Transport (DOT), Marine Safety Division, before operations within British waters, or for any vessel bearing a British flag. The French counterpart to the Department of Transport in matters of marine safety is the Bureau des Affaires Maritimes, and they have agreed to accept DOT certification.

To facilitate operational insurance, the vessel may also be classified. In this case the classification authority selected (with the approval of the Department of Transport) was Det Norske Veritas (DNV), whose rules for design and construction also then applied. DNV surveyors inspected both the habitat and atmospheric transfer vehicle (ATV) during construction and testing stages.

The maximum depth of the English Channel in the area of the cable routes is approximately 60 m, so a maximum working depth for design purposes was chosen as 65 m. The main pressure hull, complicated by the large opening door, was designed for this working depth in accordance with the rules of the DOT using a finite-element computer program.

The habitat with ATV attached, when on the sea bed, is subject to many currents and water movements, which tend to exert an overturning moment on the assembly. The most important of these forces is that exerted by surface waves at specified seastates. Calculations showed that the habitat could only be operated at or below some minimum depth. This figure was taken for design purposes as 20 m. Thus, for sea depths less than 20 m, the habitat cannot be used.

Because the habitat weighs approximately 100 tonnes (105 tonnes), it was designed so that it could be towed to location, thus obviating the need for a large, sea-going barge mounted crane.

Submersible craft of this type are not usually towed on the surface for any significant distance, and towing trials using a model were undertaken during the detail design phase. The seastates for deployment are as follows:

- launch/recovery in a mean significant wave height of 3 m;
- operation in conditions associated with a wind speed of force 6 on the Beaufort scale;
- survival in conditions associated with the Beaufort force 9 and a sea-bed current 3.5 knots.

With the exception of the survival condition, however, all practical operating limitations are contingent on both vessel and surface handling system performance criteria.

The habitat (Fig. 4) is a sea-bed vessel, which receives power, air

Fig. 4 The habitat.

and communications via an umbilical cable connected to a surface support vessel. It is equipped with tracks to give limited sea-bed manoeuvrability. The hull is a thermally insulated steel pressure vessel, 11 m long and 2.2 m in diameter. This consists of two separate compartments (living and working; Fig. 5), with a pressure-tight bulkhead and interconnecting door. The jointing area is 1.25 m in diameter and 7 m long, in which the repair is made.

The cylindrical jointing area is split longitudinally, the lower half consisting of a door which opens both to receive the cable ends for

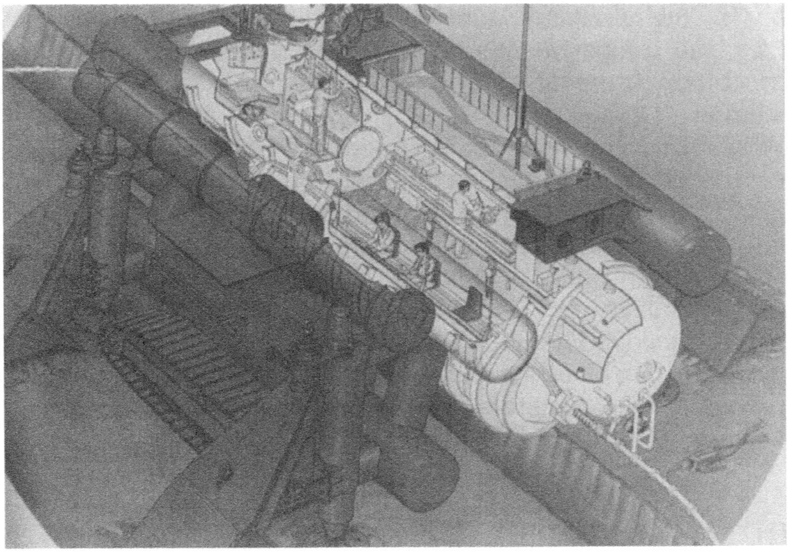

Fig. 5 Cutaway drawing of the habitat.

repair and to allow their return to the sea bed. This door is opened and closed by rams and secured in the closed position by latches, both being operated hydraulically. A specially developed annular assembly on each cable end, which fits between the jointing area door and its mating face, completes the pressure-tight seal.

THE WORKING COMPARTMENT

The working compartment is 7.8 m long, and includes the 7 m long by 1.25 m diameter jointing area. The lower half of this jointing area is formed by the half-cylindrical cable door, and at each end is an annular cable seal. The cable door is hinged to allow it to drop down and the cables to be hoisted into position. Two large hydraulic rams, controlled by divers, open and close the cable door as required.

Once closed, mechanical latches – again hydraulic and controlled by divers – pull the cable door on to its O-ring seals, thus forming a pressure-tight seal.

Within the working compartment, above the centreline of the jointing area, is a lifting beam equipped with chain hoists which the divers use to position the cable. Sliding seats and powerful lights are installed to aid the cable repair crew in their task. Storage is provided for all materials – including scrap material – and tools (both power and hand) required, this being arranged to provide complete protection from the high pressure, salty and humid atmosphere which will exist during the compartment's flooded phase. Video cameras allow the surface crew to monitor progress of the repair, and also the crew's condition.

Whilst both the living compartment and the working compartment have the same basic environmental control requirements – that is, to support human life – in the working compartment facilities exist to control the extra problems imposed by the cable-jointing process. This process demands enhanced dehumidification to achieve the relative humidity of 20%, but also produces carbon monoxide and nitrogen oxides. The use of cleaning solvents (trichloroethylene) is also taken into account. Installations to ensure that the specified environmental parameters are met are built into the compartment, being powered from the 120 V and 24 V d.c. system.

THE LIVING COMPARTMENT

The living compartment of the habitat, where the pilot will stay during the work, is 3 m long and is shown in Fig. 5. In the roof of this compartment, and reached by a short portable ladder, is the main entrance hatch, above which is the access trunk for the ATV.

In addition to environmental-control facilities and video cameras, this compartment has two bunks, a chemical toilet, a washbasin and microwave oven to enable the crew to be relatively comfortable during their stay.

From the living compartment, the pilot monitors and controls many of the habitat systems. Because the living compartment always has a dry 1 atm environment, these systems are designed accordingly.

HULL SUPPORT STRUCTURE

The hull is suspended from a plate-steel bridging structure on which are mounted four legs and the tracks, both hydraulically operated. On top of the structure is mounted the umbilical connection which supplies power at 1000 V 3-phase from the surface for all of the systems. Built into the structure and forming the main fore and aft beams are four cylindrical tanks, the upper ones being for ballast, each being subdivided, and the lower two containing the batteries. These batteries, in the event of an emergency, provide 120 V and 24 V d.c. The normal supplies via the umbilical cable are fed from transformers in a watertight pressure-compensated housing on the outboard battery tank. Adjacent, and controlled by divers from a control panel forward on the starboard side, is the hydraulic power pack which supplies all the requirements of the habitat. Large compressed-air bottles for ballast control are mounted on the port side. The oxygen bottles, however, are mounted externally on the hull itself. Video cameras in universal mountings allow the operation to be viewed from the surface.

THE ATMOSPHERIC TRANSFER VEHICLE

The atmospheric transfer vehicle (ATV; Fig. 6) is a steel cylindrical pressure vessel with dished ends, 1.85 m in diameter and 2.3 m high, weighing 8 tonnes and equipped to transfer the five-man crew

Fig. 6 The atmospheric transfer vehicle.

between the surface and the sea bed. The ATV is handled by a lift wire and run on to the habitat by guide wires installed between the mating skirt of the habitat and the surface.

A steel frame protects the ATV hull and the equipment attached to its exterior; it also acts as a support structure for some of the heavier parts. An umbilical cable supplies only power and communications, all other facilities for normal and emergency operation being carried on the ATV. On the lower part is mounted the drop weight and its four winches which allow automatic latching with the habitat mating skirt. Around the circumference are mounted the oxygen and air bottles with their associated control panels, electrical distribution, hydraulic power pack, dewatering pump and closed circuit television cameras.

Internally there is a circular bench with five seating positions, one being adjacent to the control and communications panel for the pilot. The ATV is also fitted with internal lighting and viewports.

In an emergency the ATV will make a free ascent, being equipped with manually operated guillotines to cut the lift wire, guide wires, drop weight winch control wires, drop weight latch hydraulic hoses and the umbilical.

Batteries under the seating supply on-board power requirements and through-water communications. Oxygen breathing masks are installed as a backup to the normal systems. Once the ATV has jettisoned the drop weight, it is positively buoyant and will rise to the surface.

CABLE SEALS

The most critical aspect of the design and operation of the habitat is the cable sealing, which had to produce a safe and repeatable pressure-tight seal. Although the original study recommended cryogenics, ACB/SEL, in conjunction with the CEGB/EdF, agreed that other methods should be studied. As a result the seal now used employs a steel cylindrical sleeve to form a two-stage seal as shown in Fig. 7.

The sleeve is a close fit in the circular hole formed by the jointing area wall and the cable door. On this external surface are O-rings interspersed with circumferential grooves. The seal is effected by pumping grease at high pressure via the grooves into the annular space between the O-rings, thus preventing water ingress. The sleeve fits over the polythene cable sheath, leaving an annular gap of approximately 6 mm. This annular space, at both ends of which is a flange and O-ring seal to the cable sheath, is filled with a semi-flexible

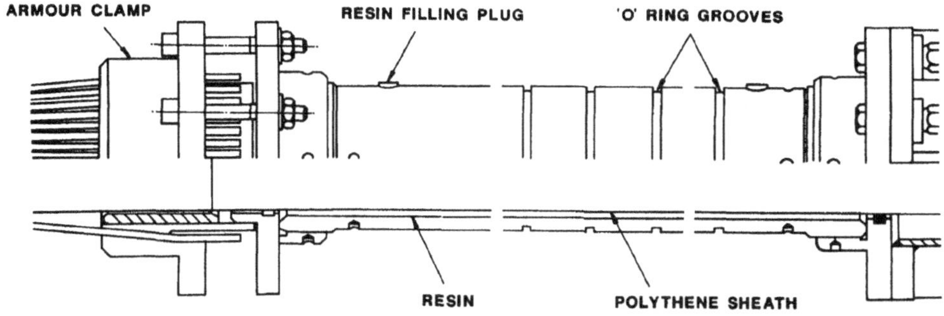

Fig. 7 Two-stage seal using steel cylindrical sleeve.

cold-setting resin, which completes the pressure-tight seal.

The armour wires are cut back and placed in a clamp which forms one end of the sleeve, the opposite end being fitted with a flange and O-ring on to the polythene cable sheath. A water-tight rigid cap and pulling head are fixed to the cable end to allow handling without cable damage. A special semi-flexible cold-setting resin is then introduced via injection points into the annular space between the steel sleeve and polythene cable sheath, and this is allowed to set.

Since the sleeve cannot be subsequently removed, it remains on the cable, and becomes part of the mechanical protection of the joint, the armouring wires being terminated on the outboard end in a specially designed clamp. After completion of the cable joint the two seals are joined together with a spilt steel shell, making the whole assembly rigid.

Because of the unique nature of the cable seal and its vulnerability, a special series of trials were used to prove its safety and effectiveness.

After curing, the resin was subjected to temperature cycles from 0 to 60°C, including a rapid cooling to 0°C to simulate immersion in sea water following exposure to the sun on a ship's deck. It was then alternately pulled in either direction with a force of 1 tonne to simulate any possible accidents. Finally, the seal was required to remain leak-tight with an internal water pressure of 10 bars, which represents a safety factor over the operating depth of 65 m.

The cable repair habitat is only one part of a complex repair procedure involving fault location, cable location, deburial, cutting, jointing and subsequent reburial. This procedure is set out below in some detail.

FAULT LOCATION

Conventional methods of cable fault location (ICE – impulse current echo) are used with an accuracy of location to ± 100 m is achievable and acceptable

CABLE LOCATION

During trenching and laying operations a very accurate (5 m) navigation system was used and UTM co-ordinates from this data are used to relocate the cable route. Following electrical location from land, faults are pinpointed using an electrical discharge and hydrophones.

DEBURIAL

A specially designed sea-bed machine is used to excavate the trench and expose the cables for some 200 m each side of the fault position.

CUTTING

A purpose-built frame is used to cut the faulty cable and lift it to the sea-bed level where equipment can be attached to lift the cables to the surface.

The procedure for a complete repair will follow, in the main, the methods used in the surface repair described earlier. The difference will be in the final joint which, instead of being made on a surface vessel resulting in a bight of cable being laid on the sea bed, will be made in the habitat.

One end of the cable will be lifted to the surface, cut back to good dry cable, tested and prepared for jointing on the sea bed, including installing a habitat seal on the cable. This end is then returned to the trench.

The second cut end is then brought to the surface, cut back and tested as above. A new length of cable is joined on the surface to the tested cable, in the normal way, sealed and lowered back into the trench until it overlaps the first end.

This second end will then lie alongside the previously relaid cable and a jig will be used to mark the position required for the habitat seal. The cable will then be returned to the surface, the seal placed in the correct position, the cable cut to length, capped and returned to the sea-bed trench.

The habitat will then be placed on the sea bed over the trench in the correct position to receive the cables for jointing. It manoeuvres into position over the prepared cable ends. Divers then carefully lift the ends through the open door into the flooded work compartment. The cable seals are positioned at their correct locations, the door is closed and latched, and the sealing system is pumped up to complete the pressure-tight seal around the door.

Using an internal dewatering pump and discharge arrangement, the work compartment is then pumped dry. At the same time the internal pressure is reduced from sea-bed ambient to atmospheric, and ventilation started via the umbilical, with air from the surface.

The atmospheric transfer vehicle (ATV), which is a separate steel-hulled five-man vessel, is then used to transfer the crew to the habitat. Guide wires, down which the ATV travels, are fixed between the habitat hatch and the surface support ship. Using a dropweight, the ATV is mated with the outboard end of the hatchway trunking to form a seal. The hatchway trunking is then pumped dry and the pressure reduced to atmospheric, using a pump mounted in the ATV, thereby forming a vacuum seal. The crew then transfers to the living compartment of the habitat. The pilot ensures that the working compartment is safe, and the repair crew enter to perform the joint.

During the performance of the cable joint, communications, life support and environmental controls are constantly monitored by the ATV pilot and the supervisor on the surface.

The following data are continuously indicated:

- habitat internal and external pressures;
- carbon monoxide and carbon dioxide partial pressures;
- oxygen partial pressure;
- relative humidity;
- internal and external temperatures;
- gases-scrubbing function.

The duration of the life-support functions with a full crew is 72 h with a further 100 h for use in an emergency. Whilst normal communication is hard-wire via the umbilical, there is a backup system of through-water transponder/receiver.

Jointing of the cable, once commenced, is continuous to minimize any contamination by moisture or foreign bodies. Since a single joint will take between 36 and 72 h to complete, there will be at least two jointing teams working in shifts. Two possibilities exist for ensuring this method of working:

- the "off-duty" repair crew retire to the living compartment and rest in the bunks provided; the habitat and ATV staying together on the sea bed for the duration of the repair;
- at every shift change, the habitat is evacuated, the ATV returned to the surface, and a fresh crew transferred to the habitat to continue the jointing process.

Either method will be possible. The separate cable contractors involved favour different methods, so both will be used. Whichever method is adopted, however, it is required to plan for foreseeable emergency situations. The most obvious problem is the fact that the repair crew are not trained divers or submersible pilots; thus, if the one pilot is incapacitated in any way, there must be contingency plan for safe evacuation of the habitat, and return to the surface of the ATV.

After completion of the cable joint, the repair crew leave the work compartment, closing and sealing the interconnecting door. The whole five-man crew then enter the ATV, closing both the habitat and the ATV hatches.

Under the pilot's control, the trunking is flooded via an inlet valve and the seal released, allowing the ATV to be winched to the surface. During the ascent, the dropweight is recovered using winches on the ATV.

With the crew on the surface support ship the habitat work compartment is pressurized to ambient, and controlled flooding allowed to take place via an inlet valve. The cable door is unlatched, jacked open and the repaired cable carefully returned to the trench by divers using pulling tools and anchor weights.

As stated previously, men capable of performing the cable joint are highly skilled technicians, usually of mature age. It is further assumed that they have no experience in diving or submersibles. The internal controls of the habitat and ATV will probably represent the most complex piece of machinery that they have ever encountered, a fact which some of them may find frightening.

It is necessary to select volunteers only, since the working conditions will be outside their usual experience. Further, an introductory period is allowed to familiarize them with the working environment of the habitat.

A further complication arises by virtue of the international nature of the venture. The native language of the repair crew may be

French or English, that of the pilot and supervisor either French or English. It is paramount that all parties are able to communicate adequately, not only under normal circumstances, but also in any foreseeable emergency situation.

Since both the habitat and ATV are complex devices, some training of the repair crew was considered essential. In addition, because of the inherently dangerous nature of the undertaken, a strict discipline must be accepted by members of the repair crew. Once the repair teams are selected, instruction in the operation of both habitat and ATV is carried out in conjunction with the pilots and divers. This training was carried out in the factory with the habitat dry, and included the making of complete cable joints. The crew were trained in both normal and emergency procedures, so that confidence was gained before embarking on "wet" trials, in a flooded dock. During this period, the divers learned how the habitat is handled on the sea bed.

TESTING AND TRIALS PROGRAMME

The two utilities, EdF and CEGB, have little experience in the testing and operation of submersible craft. It was considered that the input from the two design authorities (DNV and DOT) combined with the contractor's expertise and the utilities' knowledge, would form the basis of a full test programme. As stated previously, the contractor is a joint venture project set up by separate French and English manufacturers, thus the expertise and experience of both are available.

The overall testing schedule had four distinct phases:

(1) Proof and certification testing, during which components of the system are tested against acceptance criteria. Examples are the habitat hull and ATV which are tested to 1.3 times working depth, and the seals which are tested to 10 bars.

(2) Function testing to demonstrate that all subsystems operate effectively within their specified or design limits.

(3) Factory and dock trials. A repair crew to perform a cable joint in the habitat whilst it is still in the factory. This phase was not only used to test all functions together in a working situation, but also to familiarize the repair crews, ATV pilot, divers and support staff with the habitat. A second part of this section of the trials took place in a flooded dock, to test further the systems underwater, and train the total repair crew.

(4) Sea trials. The habitat was towed to location and deployed on the sea bed to perform a joint. This was a full trial, with all systems satisfactorily tested and a joint made on a submarine cable in the English Channel.

EMERGENCY PROCEDURES

During the preliminary trials emergency procedures were rehearsed, and equipment was tested to ensure that as little as possible is left to chance. Endurance of the emergency power and life-support systems were proven, as well as escape procedures. These were demonstrated during the trials and were proved effective.

EdF and CEGB have agreed that it is probably not economic to retain the services of a dedicated cable support ship. However, for the first year it has been agreed that one of the cable-installation support vessels will be retained to support any habitat operations. This facility may be extended for 1 or 2 years depending upon the number of faults that occur during this period. It is anticipated that a "vessel of chance" will be used in subsequent years if the incidence of faults is very low. Such a vessel would be fitted out as required and mobilized for a specific repair. Further, for the anticipated number of repairs, it was decided that only one facility would be required. Thus, only one sea-bed habitat has been built, together with all auxiliary plant necessary to carry out a repair operation.

GPSR Compliance

The European Union's (EU) General Product Safety Regulation (GPSR) is a set of rules that requires consumer products to be safe and our obligations to ensure this.

If you have any concerns about our products, you can contact us on ProductSafety@springernature.com

In case Publisher is established outside the EU, the EU authorized representative is:

Springer Nature Customer Service Center GmbH
Europaplatz 3
69115 Heidelberg, Germany

Batch number: 10761599

Printed by Printforce, the Netherlands